Bright Boys

Bright Boys

Tom Green

A K Peters, Ltd.
Natick, Massachusetts

Editorial, Sales, and Customer Service Office

A K Peters, Ltd.
5 Commonwealth Road, Suite 2C
Natick, MA 01760
www.akpeters.com

Cover image courtesy of MITRE Corporation.

Library of Congress Cataloging-in-Publication Data
    Green, Thomas J.
    Bright boys / Tom Green.
        p. cm.
    Includes bibliographical references and index.
    ISBN 978-1-56881-476-6 (alk. paper)
    1. Air defenses–United States–History. 2. Electronics in military engineering–United States–History. 3. Electronic digital computers–History. 4. Information technology–History. 5. Computer engineers–United States–Biography. I. Title.
    UG730.G74 2010
    623′.043097309045–dc22

                                                    2009037048

Printed in the United States of America

14 13 12 11 10   10 9 8 7 6 5 4 3 2 1

Dedicated to all those bright boys past, present, and future who dare to dream, to take a chance, to explore, and thereby to discover. As James Bryant Conant put it: "Behold the turtle. He makes progress only when he sticks his neck out."

Contents

Foreword

This book tells the story of the Barta Building and the external world from which the Whirlwind computer emerged. It is an authoritative account about a team of individuals who pioneered technology that forever changed our world. But there is more to the story—the internal MIT environment that fostered innovation. Especially in the 1940s, MIT can be described as a free-enterprise society in which one could do anything if it was honorable and one could raise the necessary money from the outside. In this freedom of action, I believe MIT differed from almost all other academic institutions.

Such freedom to innovate resided in the Division of Industrial Cooperation (DIC) directed by Nathaniel Sage. The DIC had been set up in the 1920s to handle contracts and funds, mostly from industrial companies who wanted to sponsor specific research projects. The DIC then became the contracting vehicle for government-sponsored activities during and after World War II. The DIC was parallel to and mostly independent of the MIT academic departments and could independently hire people.

Nat Sage set the tone of the Division of Industrial Cooperation. He favored results over the restraints of contracts and bureaucracy. As the son of an Army officer, he grew up in Army camps around the world. Somewhere in that experience, Sage became a very good and self-confident judge of people. There were those at MIT whom he trusted implicitly, and those whom he would not trust farther than he could watch them. Sage trusted Gordon Brown, my mentor, and Stark Draper, famous for gyro computing gun sights in World War II and the guidance system for the moon landing. I was also on his favored list, and he was a tower of strength in tough times, of which we had many.

An incident in the later stages of designing the SAGE computers showed Nat Sage's way of dealing on our behalf with distractions. We had several engineers who had spent a stressful week in Poughkeepsie, NY, pushing the IBM engineers to design for the reliability to which we were committed. Came Friday afternoon, they wanted to be home for the weekend, and rather than face the tedious trip to New York and a change of stations for a train to Boston, they chartered a plane to take them home. When the Air Force finance officer saw the charge for the plane, he came to Nat Sage to berate us and demand some kind of punishment. Sage was sympathetic, said he would look into it, and after the officer left put the papers in his desk. Several months later he mentioned the officer's visit to me.

Unlike most organizations, in the DIC bad news flowed upward. If there was no news, those at upper levels assumed matters were under control. The upper levels were there to assist when there was trouble, and when necessary we took problems "up stairs" for help. Also, administrative levels were not barriers to communication; everyone was free to skip levels, both up and down, to go directly to the place where discussion could be most effective.

The core team that created Whirlwind emerged from the early 1940s in the MIT Servomechanisms Laboratory directed by Gordon S. Brown. Brown was principally responsible for my career at MIT. Brown ran what at the time seemed like a chaotic organization. World War II projects came in; he turned most over to graduate students to manage, keeping a few projects for himself to manage. We, as employees of the boss, had to compete with him for resources and influence. Brown was impatient, with little sympathy for incompetence. He would "walk on" those who did not defend their positions, but respected and gave tremendous help to those who would stand up to him and demonstrate justifiable independence. I think my turning point with him came one day when I had an issue to discuss with him and went to his office door where he was sitting at a desk facing the door. I stood in the door waiting to be invited in and was sure he knew I was there. After a long time he looked up and said, "Get out!" At that point I walked to his desk, sat on the paper he was writing, and told him I would not leave until he paid attention. It was a rough environment but powerful in character building.

In developing military equipment, we gained experience in the total sequence of new products—defining the objective, innovating new ideas, studying the theory of operation, making demonstration models in the machine shop, designing for production, devising tools for factory manufacture, solving factory production problems when they arose, and going to the military field to analyze and correct failures in operation. When we started such work, MIT was without guards and military security; to fill in, I was given a City of Cambridge police badge and a pistol permit to double as a guard. At age 24, I signed drawings in the space reserved for the Chief of Army Ordnance and sent them directly to factories; such was the freedom and urgency of wartime. Here we came to a deep and abiding faith in Murphy's Law that, if anything can go wrong, it will. That enduring belief served us well as we embarked on designing reliable computers. The Servomechanisms Laboratory under Gordon Brown trained innovators.

Today, many critics lament the lack of innovation in our society and draw the conclusion that more emphasis on teaching mathematics and science will lead to innovation. That will probably fail. Pressures in the present school systems suppress innovation. Innovation comes from re-peated successes in innovating. Innovation means trying ideas outside of the accepted pattern. It means providing the opportunity to fail as a learn-ing experience rather than as an embarrassment. It means living part of the time outside of the traditionally accepted track. An innovative spirit requires years for developing the courage to be different and calibrating oneself to identify the effective region for innovation that lies between the mundane and the impossible. Almost none of the conditions for develop-ing innovative attitudes are to be found in today's K–12 education. In fact, the traditional school powerfully suppresses any tendency toward being innovative. Both teachers and students are driven to conform.

However, today's educational shortcomings did not exist in Brown's laboratory. Each project opened into uncharted territory. Each provided a wide range of experience that far exceeded that of most people in industry, government, and academia. It was this wealth of experience and building the personal character for innovation and entrepreneurship that produced the team that was capable of pioneering the digital frontier.

Gordon Brown had an even more crucial role by launching the activ-ity that led to the Whirlwind and SAGE programs. The end of World War II terminated the military projects in the Servomechanisms Laboratory. I assumed that I would be leaving MIT either to find an industrial position or to start a new company to carry forward and apply our feedback system knowledge to physical engineering applications. Then, Gordon called me to his office to ask if I would like to stay and choose one of the ten projects that he had on a list.

I decided to pick one from the list. Captain Louis de Florez at the Navy's Special Devices Center in Port Washington, Long Island, had proposed it. De Florez was an uninhibited innovator. One story from his early career comes from when he worked for an oil company in Texas. Refineries were threatened by fire and explosion when a pipe eroded from the inside until the wall was weakened, and a major bursting and explosion could occur. De Florez devised a warning system by drilling small holes part way into the pipe walls. When the inside eroded to the bottom of a hole, oil would spray out as a warning. De Florez was a commanding figure with a pointed waxed mustache. Also, as far as I know, he was the only person able to get permission to land a seaplane in the Charles River sailing basin in front of MIT. I was there once when the Metropolitan District Commission cleared the basin of sailboats; de Florez landed his plane and came to the MIT alumni reunion. After lunch, when the speeches were becoming dull, he left in a roar from his water takeoff that overwhelmed the loud speakers at the reunion.

The de Florez project was to go beyond the existing flight trainers for pilots, which were tailored to the characteristics of a known airplane. The project was to create a cockpit that would exhibit the characteristics of a proposed plane, based on wind tunnel data of a model of the plane. Thus started the sequence described in this book.

Although the Division of Industrial Cooperation was outside of and parallel to the MIT academic departments, projects did have a tenuous connection to a relevant department. Almost always, there was a faculty member who bridged between a DIC project and an academic department. Gordon Brown played the role of connecting the digital computer activity to the MIT Electrical Engineering Department.

We did have frictions in trying to awaken a traditional electrical engineering department, devoted to power generation, to the arrival of computers. At one point we presented, with only partial success, a lecture trying to convince the electrical engineering faculty that it was not only possible but also desirable to use binary arithmetic for computation.

Our relationship to the Electrical Engineering Department gave me access to incoming graduate student applicants to become full-time research assistants. They were allowed to take two subjects per term toward their degree. By reviewing applications myself, I effectively had first choice of those applying to MIT. We accumulated an impressive group. Many graduate applicants were coming back from military service with a maturity and real-world experience that is unusual for students. They were ideal recruits for the coming information age. Their M.Sc. theses on aspects of computers qualified many to present papers at the national and international technical conventions.

Because of the continuing budget pressures in the early phases of the Whirlwind computer, we were subjected to annual reviews by panels appointed by the Office of Naval Research. Most of these were a distraction from our mission, but occasionally a point was raised that improved our thinking. An especially important advance came when ONR appointed a mathematics professor, Francis Murray of Columbia University, to be inquisitor-of-the-year. Murray came on a Saturday and we met in my office.

At one point he asked, "What are you going to do about the electronic components that are drifting gradually and are on the edge of causing mistakes? Any little random fluctuation in power, or streetcars going by, will cause circuits to sometimes work and sometimes not." This was a very perceptive and powerful question. Inexplicably, we had done nothing about it. It was such a pointed question, and obviously such an important one, that I felt an immediate answer was essential. I said to him, "Well, we could lower the voltage on the screen grids of tubes to change their gain and convert behavior from a marginal to a permanent failure and then it would be easy to find." He thought it was a good solution and so did we. The next week we started designing such capability into Whirlwind. The "marginal checking" system in Whirlwind carried over into the SAGE Air Defense system, adding another factor of ten to the reliability.

During the design of Whirlwind, we were strongly criticized for the amount of money we were spending to develop entirely new circuits and devices that would be reliable. The total Whirlwind cost over some seven years was four and a half million dollars; it would not be long after until individual production line machines were costing that much. The matter of cost was one of the things that the outside world understood least. Whirlwind was being judged in the context of mathematical research, in which the salary of a professor and a research assistant was the standard by which projects were measured. We were spending way beyond that level and were seen as running a "gold-plated operation." Although the gold plating might occasionally have been excessive, in retrospect, I think there was reason for it.

An organization has great difficulty maintaining two contradictory standards. If you're going to have high performance and high quality in the things that matter, it is very difficult to have low quality and low performance in the things that, perhaps, don't matter. The two standards bleed into one another to the detriment of both. For example, at an early demonstration for important people, we didn't want them sticking their fingers into the high voltages in the circuit racks of Whirlwind. I asked somebody to get rope to put along the aisles so visitors wouldn't walk among the vacuum tubes. A nice-looking white nylon rope was procured and installed. During the demonstration, I saw some of our critics fingering this beautiful rope and looking at one another knowingly as if to say, "That's what you would expect here." It may not have cost any more than hemp rope, but it reinforced that impression of an extravagant operation. Another example was the Cape Cod display scopes built into cabinets made of mahogany-faced plywood. Although our cabinetmaker made these quite inexpensively, people looking at those mahogany cabinets were reinforced in their thinking that we were extravagant. Eventually we solved this problem by spending additional money and painting the cabinets gray.

We came out of the Servomechanisms Laboratory with a central group that understood the full potential of teamwork. Of course, later, the team unity was reinforced as we stood together on the computer frontier to repel critics. There was complete sharing of information. In a bi-weekly report, distributed to everyone, each person reported on progress and difficulties. There was little jockeying for personal advantage. Every person has strengths and weaknesses. A team must have a shared vision of the future, a sensitivity to political matters, the capability of developing people, technical competence, the courage to transcend adversity, salesmanship, integrity, the ability to put long-range goals ahead of the short term, and a shared understanding of the individual strengths and weaknesses within the group. We had those characteristics well represented, scattered throughout our group. No person had all these skills. For every person there would be a glaring hole in one or more of those dimensions.

Yet, it was a group that understood each other well enough to use people in situations where their strengths prevailed and have others compensate for their weaknesses. Out of that came an organization that was able to be much more effective than most of those we see around us in technology and corporations at the present time. It was an organization possessing power based on a clear vision and consistency that could dominate the military and corporate structures that grew around the North American air defense system.

Jay W. Forrester
September 15, 2009

Chapter One
On the Road to Find Out

You can't really guess where mind-amplifying technology is going unless you understand where it came from.

—Howard Rheingold, *Tools for Thought*

Bright Boys is a story about technology when technology was young: 1938–1958. Two decades that ushered in the new world of electronics. A remarkable 20 years when computers were giants, their makers young and unknown, and when there was less than a megabyte of random access memory on the entire planet. Microprocessors were science fiction; transistors were handmade and mistrusted; and banks of hot, glowing electron tubes ruled the land. Networks were meant only for telephones and electric power grids, and the word "digital" was new on the ear. Yet it came as no surprise to a band of bright boys that these two decades would change everything and that they would play an enormous part in getting it all started.

Their first decade was filled with guesses, probings, and dawnings. The second was a near-magical time of incredible emergences as, almost simultaneously, information theory came of age, digital computing first appeared, networking was born, and the first glimmerings of Information Technology edged into existence. Astonishing was the convergence of information, digital computing, and communications wrought by a bunch of cocky, twentysomething engineers, who thought they could do most anything they turned their hands to, and usually did. Others contributed mightily to the convergence as well: brilliant aviators from the Air Corps Tactical School at Maxwell Field, Alabama; radar pioneers from Ft. Monmouth, New Jersey; and physicists from MIT's wartime

Interior of Barta Building and Whirlwind computer; Jay Forrester with Norman Taylor. (Photo courtesy of MITRE Corporation.)

Radiation Laboratory. Theirs was an amazing bucket brigade of bright boys handing along their intellectual toil from one to the other, from one decade into the next. Equally amazing and improbable was that these same bright boys would themselves eventually converge, gathering together with their ideas and energy in the most unlikely of places: an old, rundown former laundry building in Cambridge, Massachusetts, a few hundred yards from the banks of the Charles River.

The Charles is at its widest where it separates the city of Boston from that of Cambridge. Massachusetts Avenue and its bridge—known locally by the same name, but actually Harvard Bridge—link both. On the Cambridge side, a stone's throw from the front entrance of MIT, is 211 Massachusetts Avenue, an ornate, hundred-year-old, brick building that when first erected in 1904 was known as the E&R Laundry Company, and then, much later, called the Barta Building, after a publisher by the same name. From the bare floor of that vacant factory the bright boys launched the country on one of the greatest and most successful projects in the history of American engineering. No similar enterprise had ever or has since sprung from such humble surroundings.

If they had brought forth only the first real-time digital computer, the first digital network, and the first practical application of information theory, as they did, and produced nothing thereafter, their adventure would have been a colossal triumph. However, their string of startling successes rambled on.

They were the first to develop an electronic digital computer capable of controlling other machines and processes, a distinctly radical departure from the handful of other electronic computers of the day that calculated numbers only and then regurgitated an answer. They pioneered the highly complex and very chancy process of engineering multiple electronic systems together into a single system. They were the first to institutionalize quality control from R&D to prototyping to manufacturing and on into management. They originated modular construction of computers. They invented the modem and taught AT&T how to use it. They conceived of IBM's first computer assembly line. They originated computer memory that became the industry standard until the 1970s. They launched the formal development of software programming. And like major, high-tech corporations years into the future, they designed and prototyped most everything yet manufactured next to nothing, preferring to outsource all production and supply. Theirs was a profusion of technological firsts that is as yet unmatched in any single engineering project since.

When all the bright boys first gathered together there was no telling what would transpire in that old factory when youth, intelligence, energy, and pluck were implored by their country to do the impossible, like figure out a way to defend the skies over North America from enemy bombers. By 1950, these young MIT-trained engineers and their cohorts from the United States Air Force were given that exact challenge—with the added proviso to get the job done in a hurry. Their task was especially urgent because the Soviet Union had just detonated its first-ever atomic bomb and had unveiled a fleet of long-range Tupelov bombers readily capable of delivering deadly payloads into the heartland of America. Their task was especially important in that no such air defense even remotely existed and very few had even a vague notion of how to make one. Many people, some influential, dismissed their task as impossible and an utter waste of government money. Still others, some of whom were influential believers but fearful of the new technology, inveighed mightily against their attempts to build it. In short, it was a perfect scenario for the making of the future, and the bright boys were not about to disappoint.

For sheer size, the bright boys' undertaking was rivaled only by the Manhattan Project; and for sheer innovation, its scads of brilliant inventions were for decades thereafter without peer. If the country had pinned its hopes on the Manhattan Project to end a war, its trust was pinned on the bright boys to help save a world. They not only made the impossible possible for the skies over North America, but, in the process, also ushered in the military's long sought after yet previously elusive prize, real-time command and control. And as if that wasn't enough of a success, they poked a gaping hole in the future and dragged into being the modern world of Information Technology.

Initially, the bright boys had little idea what exactly it was they dragged in through that gaping hole. Their search was for how best to manipulate digital information for air defense using their newly hatched creation, the Whirlwind computer, a monster of a machine that filled an entire floor of the Barta Building. Information Technology, a term not yet coined, was an unexpected consequence of that primary objective. Although they pioneered Information Technology, the closest they got to a name for their work was "real-time command and control," which was the information engine for how air defense went about spotting far off enemy bomber formations and getting interceptors after them. However, scratch the surface just about anywhere over the enormous swath of Information Technology—from email to file transfer to smart bombs over Baghdad—and there gleaming back up is the unmistakable DNA of real-time command and control.

Each time we spin a CD, step off an elevator, book an airplane ticket, get cash from an ATM, or even pop a steak into a microwave, there's an unseen thread that leads directly back to the bright boys' inner sanctum of discovery and innovation. Of course, now everyone and everything is bound up in the clutches of this all-pervasive, all-powerful Information Technology. Popularly known to most as just IT, Information Technology is that high-tech panoply of computers, telecommunications, and the myriad devices that digitally connect and integrate information, equipment, and people. Beginning with the bright boys' first information loom, built and cranked into action in the old factory near the Charles, these two initials have subtly yet inexorably woven themselves into the fabric of our lives.

Today there are information looms everywhere, atop our desks, on our laps, and in our pockets and purses. Many of us spend most of our waking hours spinning information on them, creating, shaping, and forming information as needed, then sending it at the speed of light everywhere and anywhere at any time to anyone of our choosing. As the *Stanford Encyclopedia of Technology* sees it, Information Technology's fingers are into everything, essentially all the "collecting, storing, encoding, processing, analyzing, transmitting, receiving, and printing of text, audio, or video information." That covers just about everything we do in today's digital world, including helping us to earn a living, raise a family, and enjoy ourselves.

We've gone so far as to define our times as an information age, an information economy, an information society, or an information revolution. Not so in the world that swirled around 211 Mass Ave. Back then there was a pristine simplicity to the notion of information. Even as late as the mid-1940s, information was meant mostly for the likes of the US Census Bureau every ten years, banks, and insurance actuaries. There were, in all of the technical and scientific literature for 1946, only seven articles on information.[1] The word "information" seems simple enough, deriving from the Latin, *informare*, meaning, "to put into form." However, as the *Stanford Encyclopedia of Technology* makes us aware, when the word "technology" is appended to the word "information," all hell breaks loose. Today the word "information" has taken on a whole new spin: it's anything and everything… and it's powerful. Charles Seif, the well-known physicist and journalist, likens us to information beings. "Each creature of the Earth," he writes in *Decoding the Universe*, "is a creature of information; information sits at the center of our cells, and information rattles around in our brains."[2] The book's subtitle hammers home that conviction: *How the New Science of Information is Explaining Everything in the Cosmos, from Our Brains to Black Holes.* No one in 1946 could ever have fathomed how transmuted information would become when mixed in the crucible of technology. The bright boys were as startled as any.

Technology in a Foxhole

If the civilian world was clueless about information, the world of the military was not far behind. Information needs for the military were immediate and vital, and it was agonizingly slow in recognizing the emerging information paradigm. Anyone who has ever peered over the top of a foxhole toward enemy lines has a healthy reverence for information. What's the enemy up to? Will they attack? If so, when and where? What are their numbers? What kinds of weapons do they have? Can we match up with them? Can we beat them? The military has an unending supply of questions about an adversary's everything that all need quick, accurate answers.

It's a natural reaction considering that outcomes like injury, capture, or death can result from bad or slow or nonexistent information. The problem with military information was that it hardly ever arrived in real time, was nearly always fragmentary, and of inconsistent quality. Without solid information with which to analyze a potential threat or to get a clear understanding of what's happening on a battlefield, victory and lives were always in great jeopardy. For millennia the struggle to gain an information advantage over an enemy was as desperate as the fighting itself.

Bright Boys is also a story about how the military got savvy about real-time information and how to acquire it. It was the United States Air Force that first stumbled upon the great teacher in the old laundry building and came to realize the information potential of the huge machine. That encounter sparked a revolution in military affairs known as real-time command and control. It transformed the Air Force from a staunch computer disbeliever "to the leading advocate of computerization in the military."[3] But the transformation did not come easily. Although code breaking and espionage were tangible enough information sources, it was difficult for the military to comprehend that an information advantage might actually be a weapon—maybe even the ultimate of weapons.

Weapons were supposed to be things that one held or rode or loaded with shells. A B-17 glinting in the sunlight and bristling with power was a real weapon, or so they thought. A couple of bright boy aviators in the late 1930s, information zealots who challenged an engrained "bomber mafia," paid dearly for their stubborn belief in the primacy of information over cold steel. Years later, one of them would eventually struggle his way into the old laundry building to finally prove the point. Time and technology have now shown the military how right the two zealots were.

"Information technology," Bruce Berkowitz, the former CIA analyst, hailed in his groundbreaking *The New Face of War*, "has become so important in defining military power that it overwhelms almost everything else."[4] A belief now held fast by none other than the United States Department of Defense as "central to greater battlefield awareness, enabling our forces to acquire large amounts of information, analyze it quickly, and communicate it to multiple users simultaneously for coordinated and precise action."[5]

In order to get a leg up on an adversary, armies have for millennia sought out two kinds of information. Strategic information is gained well before any guns go off about the intentions or plans that an enemy is hatching. Learning the intentions of the Japanese high command before the Imperial fleet set sail for Pearl Harbor would have been of great strategic value. On the other hand, tactical information is more immediate. It's early warning, like finding out in the nick of time that Admiral Yamomoto's dive bombers were revving their engines on carrier flight decks. Good, real-time information, what the military calls "actionable information," can mean the difference between victory and defeat. In 490 BC Phidippides ran 26 miles to Athens to carry tactical information of the Greek victory at Marathon and to warn about Persian ships approaching Athens. Sun Tzu, the Wu Dynasty's very winning general from the sixth century BC, had his information-gathering techniques put into book form, *The Art of War*. Knowing his enemy was all-important to Sun Tzu, and he went to extraordinary lengths to gather both strategic and tactical information. To save wear and tear on himself and his army, Sun Tzu believed in the power of gaining advantage through strategic information. *The Art of War* devotes a chapter to spies, double agents, espionage, and counterespionage.

It wasn't until the unimaginable carnage of World War I that militaries fell upon a way to get a fast, consistent, and dependable information edge on their enemies. A stalemate between opposing barbed-wire trenches for years on end with new weapons of war like machine guns, tanks, poison gas shells, lightweight mortars, and long-rage artillery taking an hourly toll of each nation's best was the agent of change. Flimsy biplanes with on-board cameras snapping pictures of battlefields provided the information breakthrough. A picture is worth a thousand words, and an aerial photo of enemy positions might well save thousands of lives . . . and maybe even win a war. The flimsy biplanes gave way to the rise of air power and the flimsy cameras to the rise of aerial reconnaissance. Suddenly planes got faster, war got faster, and pictures alone were not enough.

Code breaking was faster still and very stealthy. Breaking the Japanese naval cipher JN-25 ensured victory at the Battle of Midway, as did similar success with the German's Enigma code help to shorten the war in Europe. Both successes came courtesy of a machine: a whirring, clinking, gear-laden analog device that did not drop bombs or spew out rounds of bullets, rather, it spewed out information. And the information it produced was powerful: it changed events; it changed history. How powerful an information weapon was it to know in advance an enemy's every move? Heady stuff indeed for any general to command. Yet somehow these clanking beasts of information never evolved up to the next level. They were limited in their capabilities. They seemed stunted from birth. Something else was needed.

The Something Else

In the faint pre-dawn glow of the Digital Age—the mid- to late-1940s—electronic digital computers had much in common with Jurassic dinosaurs: large in body but very short on brainpower. Staggering behemoths of hardware, most were about the size of a gymnasium, yet with barely a fraction of the capability of a modern desktop PC. Hardly anyone gave much thought to an electronic digital computer doing anything more than munching huge quantities of numbers and then slowly regurgitating an answer. Hardly anyone, that is, except for a cocky bunch of bright boys.

Preceding the bright boys by nearly a decade and a half was the military's ongoing pursuit of improving air command and control: the ability to detect enemy planes and then to vector fighters to intercept them. In an age of jets and nuclear threat, it had grown too slow and cumbersome a process to be effective. To exert command and control in real time—as events were actually taking place—was the key objective. First manifesting itself in the minds of two Army Air Corps officers in 1935, command and control's 15-year quest for speed and accuracy would converge with the development of air defense in the old laundry building.

It's a saga of connections and crosscurrents, people and events crisscrossing through recent history. The accretion of events, roughly from 1938 to 1958, at once political, military, scientific, technological, and socioeconomic, begat its very own culture—a real-time command and control culture—which when it reached critical mass spawned a revolution in engineering. Playing out roles are air battles from the Spanish Civil War, B-17 raids over Europe during World War II, brilliant but obscure engineers in research labs, radar and radios, digital computing, information theory, cybernetics, and a fate-filled meeting in 1949 at a Pentagon taxi stand. Through grit, hard work, a little derring-do, pure genius, and out-and-out good luck, this emergent culture forged an elusive idea into a technological marvel. The pursuit of air defense would bring forth real-time command and control, which, in turn, would become the undisputed lion in all of military affairs and permeate every facet of modern Information Technology.

Air defense began more like an ever-expanding idea than a buttoned-down, ready-to-go engineering plan. The end purpose of all the putting forth was the supposed creation of a real-time, electronic command and control system capable of ensuring the detection, tracking, and interception of any number of incoming enemy bombers anywhere and at any time over the skies of the continental United States. A tall order. Many thought it wild-eyed fantasy more appropriate to comic books of the time like *Tom Corbett: Space Cadet*.

Even the customer, the United States Air Force, had grave reservations. The Air Force think tank, the RAND Corporation, issued a 1948 study highly dubious of air defense.[6] Caution seemed the better choice. Why bet the farm on something as fantastically speculative as what a small group of individuals had in mind? For here were a bunch of bright engineers and a cadre of beyond-loyal Air Force officers, who seriously considered yoking a brand new technology like digital computing—which was fighting for its own survival with the well-accepted analog computer—to World War II-era radar units, all connected by telephone lines over which would travel digital radar data. All of it brazenly new and heretofore never attempted.

Rising up and over the storm and stress from such an engineering adventure, a command and control culture began to emerge. A culture developed that integrated both the internal events of the technology with the history of the external forces acting upon that technology—what Donald MacKenzie in *Inventing Accuracy* terms "heterogeneous engineering,"[7] whereby a technological advance cannot be isolated from its social circumstances. The social and physical worlds were engineered together, giving rise to a new set of ideas, skills, and knowledge as well as devices. An original bright boy, who much later became president of the MITRE Corporation, Robert Everett, spent the better part of a lifetime actively engaged within the culture. He points out that the once brazenly new undertaking "was the predecessor of military C3 systems [command, control and computers], air-traffic control, and other large computer-based systems."[8] Indeed, this inchoate command and control culture would in short order bring forth technological wonders beyond all reckoning. Prefigured within its small, humble beginnings were entire future industries.

The need to engineer together a radar system over a telephone system to a digital computer system—and the scrupulous systems management necessary for such an intricate process—would pioneer a revolutionary new idea in engineering called the *systems approach*. This new method of complex systems building—and the tying of it all together through systems integration—would become absolutely vital to all subsequent large-scale engineering undertakings, especially the ICBM, space satellite, and NASA projects. And emerging with it came the culture's very own discipline, systems engineering: the quintessential competency for every well-schooled, high-tech engineer. Command and control's influence, through its development tools, people, and devices, would also answer the call to the military's search for a survivable, bombproof communications system by spawning the quest for its most compelling progeny, the Internet.

"The unique confluence of military, political, economic, and technological circumstances in the 1950s," writes Thomas Hughes in *Rescuing Prometheus*, "brought a decade of managerial and technological achievement that has lastingly influenced the character of American engineering and management."[9] The fervor of systems building, although primarily attendant on government and military projects, also tried casting its tendrils into the public sector. Large sections of Hughes' book chronicle the attempts—mostly unsuccessful—of migrating the systems approach to social and public works projects for state and local governments during the late 60s and early 70s. Intellectually, the systems approach had strong potential for rebuilding, say, urban housing developments or in devising public transportation systems. In reality, however, it turned out to be the right message, but from the wrong messenger. A systems builder working on single-focused, big engineering projects like the Atlas rocket just couldn't cotton to dealing with politicians, interest groups, community activists, citizens' committees, and all the other energy-sapping denizens of city bureaucracy.

Yet another tendril curled its way into the realm of business management. Thomas Haigh's paper, "Lost in Translation: Total Systems from War Room to Board Room, 1954–1968," points out the profound influence of systems engineering on the Management Information System (MIS) or the Total Systems Concept movement in business management beginning in the late 1960s. Haigh then traces the clear line of ascent as the Total Systems Concept movement transforms itself into that most bandied-about of today's *digi*-terms, the ubiquitously well-worn *Information Technology*— better known as IT.[10] Ads for UNIVAC computers appeared in 1965 issues of *Fortune* and *Business Week* touting "with a UNIVAC Total Management Information System . . . centralized control of decentralized operations becomes a reality." In 1948, however, the future importance of such a command and control culture and its staggering array of future spin offs were in little evidence and any potential outcome from its efforts was looked upon with great skepticism.

Fortunate too for the undertaking was that it sooner rather than later attracted a deep-pocketed, foursquare customer that believed much, risked much, and involved itself passionately with the project. The United States Air Force, belatedly disregarding the RAND warnings, finally anted up the cash and the nerve to get it all going. A very technologically oriented and enlightened military early on was the Army Air Service, and later under the great air power visionary, General Henry "Hap" Arnold, the then Air Force jumped in up to its chin for real-time command and control, eventually riding its big investment literally to the stars. As Arnold noted to General Bernard Schriever, the brilliant and indefatigable master builder of the Air Force space program, World War I had been won by brawn, World War II through superior logistics, but that any future war would be won by brains.[11] As "Shock and Awe" headlines in 2004 about the precision hell delivered on Iraq would amply show, the Air Force learned well: smart munitions and smart aircraft breed smart tactics. The "smarties" who began it all in their dowdy building in Cambridge squeezed the cloud of skepticism that loomed over them and turned it into a rainbow.

The development kernel within this emerging culture was a model for how Vannevar Bush, President Roosevelt's chief scientist and director of the Office of Scientific Research and Development (OSRD) during World War II, thought postwar research should go. Immediately following the war, he pushed for his vision of a civilian-controlled National Research Institute to provide for all of the country's needs. OSRD's own wartime successes like the proximity fuse, microwave radar, and the vaunted Manhattan Project, he argued, were stellar examples of the military and civilians working together. As the saying goes: "the atomic bomb ended the war but radar won it." Such a sentiment speaks volumes to the results of his development methodology as well as to the impact of science and engineering on the ensuing Allied victory. Bush cajoled and chided civilian engineers and scientists, military officers, and government officials into trusting one another to work together for a common goal. The results, well, they're known as V-E Day and V-J Day. However, just as OSRD's World War II successes had shown that disbelief to discovery was the natural order of great things, real-time command and control was no different, except that it lacked a world war to silence its doubters.

The making of real-time command and control, although sometimes contentious and fraught with difficulties, stayed true to the OSRD formula. Many of the young engineers working on command and control for the Air

Force knew Professor Bush from their undergraduate days at MIT. Some, as Bush protégés, had worked on his electromechanical analog computer, the Differential Analyzer. Still others had labored on Bush-supervised OSRD projects during the war. Through it, however, all came to fully realize and believe that it is not the brilliant researcher working in isolation who alone owns the cry of *Eureka!* Rather, it's a team.

The team of bright boys holding forth in their old factory did not have the desperation of a world war to drive their cause, but they did have the ever-escalating peril of postwar world events: Cold War clashes like the Berlin Airlift of 1948, the Soviet atomic bomb blast and intercontinental TU-4 bomber in 1949, the fall of China the same year, the Korean War from 1950 to 1954, and the Soviet Sapwood ICBM and then Sputnik—both in 1957. The resulting national frenzy from each event ratcheted up the hysteria enough to secure sizeable development dollars for the military.

The American public began looking skyward with great apprehension. Air raid sirens blared atop schools and public buildings in weekly tests, and school kids practiced hiding under their desks. Bomb shelters were all the rage while a spate of sci-fi movies screened a fearsome atomic future. The Air Force went to the movies with its own scary story, producing the stark, propaganda tale *In Your Defense* that opened with its narrator posing the looming threat:

> One of the most dangerous threats to our nation's security is the possibility of attack by high-speed enemy bombers armed with nuclear weapons. These bombers can strike at supersonic speeds from many directions and altitudes to confuse our defense and delay the dispatch of interceptor weapons. In a mass raid, high-speed bombers could be in on us before we determine their tracks. And then it would be too late to act. We cannot afford to take that chance.

This bright boys' project would rapidly grow—seemingly exponentially—as the winds of world events swirled tensely around the brick enclave on Massachusetts Avenue. The idea of how best to engineer real-time command and control into reality began to dominate the huge computer full time.

In September of 1948, the bright boys' youthful leaders, Jay Forrester and Robert Everett, together with two of their mates, Harris Fahnestock and Robert Nelson, put out a blockbuster report called "Forecast for Military Systems using Electronic Digital Computers." [11] It was a two-by-three foot foldout page on which was a 15-year forecast (1948–1963) of computers in the military, covering guided missile data reduction, cryptography, high-speed computer networks, interception networks, air-traffic control, industrial process control, simulation and training, and logistics. The implications of this high-tech roadmap went far beyond the military. It reached far into the future, far beyond what most thought capable of ever emerging from the technology of the late 1940s. It read more like a splashy spread of far-off predictions from an issue of *Popular Science* than serious ideas from serious engineers. In 1948, it was met with incredulity and was completely overlooked. By 1958, it was gospel. Today, it's the way of the world: Information Technology.

Chapter Two
Bomber's Moon

If they do not now accept our terms, they may expect a rain of ruin from the air the like of which has never been seen on this earth.
—President Harry S. Truman, New York Times, August 6, 1945[1]

It would take the military's newest paradigm for doing battle—air power—to recognize the other newly emerging paradigm, information. That they were both so new and fit so well together had much to do with that early recognition. As Marcel Proust said about such voyages of discovery, it's "not in seeking new landscapes but in having new eyes." The United States Air Force's new eyes saw the knack that information held for winning wars, but only after a nasty and internecine brawl.

Two aviators first broached the struggle for information advantage, and they paid dearly for their audacity. One was ostracized to China where he became a hero; the other joined the bright boys and lived to crow back in victory from atop the Barta Building. In the end, Information Technology would step out from the shadows and "information war" would become the Pentagon's newest dance partner. It all came together awkwardly as the air defense of North America and real-time command and control rumbled into place.

Terror on a massive scale and total devastation are best delivered by air. H.G. Wells, Hermann Goering, and Curtis LeMay knew it well and used it. Armadas approaching by sea are highly improbable threats as are land armies approaching from any direction. They are too slow and vulnerable, especially vulnerable from above. Any concern over the defense of the United States in the 1950s was skyward. Air defense was very much on the mind of postwar America. And why not?

Major John Marchetti, U. S. Army, 1937. (Photo courtesy of U. S. Army.)

Americans were newbies at living and working under the constant threat of nuclear holocaust. For them, the thought of a sneak attack was newly terrifying and moments away.

The vast majority, including the military, felt that the country was vulnerable. The decade-old yet horrific memories of Pearl Harbor were still hauntingly fresh, as were newsreels of V-2s falling from the sky over London, and the ultimate of terrors: mushroom clouds over Hiroshima and Nagasaki. Those attacks had all come quickly from the air, were startlingly deadly, and succeeded in forcing the eyes of the world to look skyward with dread and foreboding. Strangely, although the likely contender for such an attack and how it would be carried out were well known, not much was done in the way of defensive diligence to prevent one. Such oversight came into sharp focus on the night of April 16, 1952.[2]

A US radar installation off the coast of Alaska's Nunivak Island in the Bering Sea watched with growing alarm as unidentified aircraft streaked east-southeast toward the continental United States. Such sightings were not uncommon for the radar crew, but each alert was nonetheless always nerve wracking until a rational explanation was in hand and the all clear given. What made this sighting so significant was that 3,000 miles away, clear across the continent, Eastern Air Defense Forces on Presque Isle, Maine would spot more unidentified aircraft honing in on the US mainland. Heightening the tension even more was the haltingly slow, laborious, and convoluted transmission of information about this potential threat through and among Air Defense Command installations.

It would take better than an hour and a half for the first alert to arrive at headquarters of the Air Defense Command, Ent Air Force Base (AFB) in Colorado. The Nunivak crew made their radar finding at 8:30 p.m. Alaska Time, relayed it to Elmendorf AFB, also in Alaska, which, in turn, radioed McChord AFB in Tacoma, Washington, which then shuttled it along to Hamilton AFB in California, and finally on to Ent at about midnight Mountain Time. Ent passed the sighting along to the Royal Canadian Air Force and to other Air Defense Command installations. When Ent tried to confirm the sighting with Elmendorf, the lines of communication went dead.

The Vice Commander of the Air Defense Command at Ent, General Frederick H. Smith, was finally awakened with the news at 2:30 a.m. More than four hours had elapsed since the Nunivak sighting. Another 30 minutes later, with General Smith and his staff deep into their deliberations over the alert, word arrived from Presque Isle about the East Coast situation. At 3:10 a.m., General Smith ordered the country's first-ever nationwide

Air Defense alert. Hotlines buzzed across the country. Air Defense commander General Benjamin Chidlaw took command at Ent. General Curtis LeMay's Strategic Air Command was still on the ground; the Air Force's Tactical Air Command was also still on the ground, and both fixed and mobile antiaircraft artillery were idle. Soon the Pentagon would be alerted, the White House informed, and the nation's military cranked up into a frenzy of activity. All of which was woefully too late and extremely dangerous. There was no time for rational thought, situation analysis, or measured response. It was all knee-jerk reaction: crank up the interceptors and get airborne. The United States had been caught once again with its pants way, way down.

There was only one place in America, maybe even the world, for that matter, where such alert conditions could be speedily and fully recognized and where there was a system in place that rendered them nearly routine. Unfortunately for the Air Defense Command, that place was behind the weathered brick walls of the Barta Building in the able hands of its artful residents, the bright boys. By October of 1950, the bright boys had cranked up their Whirlwind leviathan into the world's first real-time digital computer. Six months later, in April of 1951, they demonstrated—with another world's first—how to detect, identify, and track unknown, approaching aircraft and how to plot jet fighters on courses to intercept them. And the bright boys' technology easily accomplished all of that detection and interception in "real time"—as the attacks were actually occurring—rather than long, dangerous hours later when there would be only time enough to duck and to vainly hope for a chance to retaliate. Nearly two years had passed since their highly successful first test. Why the delay in getting the bright boys' air defense system operational? Certainly pieces of the technology could have been rushed into play early on to cover such critical targets as the Hanford Atomic Energy plant in Washington, or the locks at Sault Saint Marie, Michigan, or the White House, or even the Pentagon.

It certainly was not from lack of money that the country fostered such glaring unpreparedness. In 1952, *Time* magazine reported[3] that the economy was in great shape. The United States' $350 billion in gross national product was "the greatest material outpouring in its history." With only seven percent of the world's population, "the U.S. produced 52% of the world's mechanical energy, and used it so wisely that it made 65% of the world's manufactured goods." No such wisdom was accorded the country's air defense. Although the United States Air Force had become a separate service in 1947, with a primary responsibility to defend the nation's air space, it had been demobilized, shrunk, and reorganized three times before 1952, with its Air Defense Command having endured complete elimination followed by a hasty reinstatement.

The US Army Air Defense Command (ARADCOM) in 1952 consisted of 45 battalions of antiaircraft artillery (AAA), which were thinly scattered batteries of 90 mm and 120 mm guns. ARADCOM's Nike surface-to-air missiles would not be operational until 1954.[4] World War II-vintage radars, solid equipment but very limited, were also thinly scattered in places like Nunivak and Presque Isle. There were jet interceptors available, but a war was going on in Korea and that took precedence. What meager air defense there was protecting the United States had little in the way of communications. If an attack was launched against the US mainland, any alert of it would be agonizingly slow.

For all of its material prosperity, the country was suffering from some grave adversity as well. A polio epidemic was sweeping the country, claiming in 1952 alone over 3,000 deaths from 57,000 reported cases. The Korean War was two years on with 1.8 million Americans contesting both North Korean and Chinese troops for control of the lower half of the peninsula. The war bogged down in 1952, and a distinct anti-war sentiment quickly set in. Before an armistice was reached in July of 1953, America suffered 54,000 killed, 103,000 wounded, and 8,200 missing in action.

The Democrats, after two decades in the White House, felt uneasy about the Republicans' new-look presidential candidate for 1952, General of the Army Dwight D. Eisenhower. The folly of national defense, *Time* reported, "rests wholly on the Truman Administration, which wanted visible civilian prosperity and no civilian hardships to bolster its campaign slogan: 'You never had it so good.'"[5] The military took the hit. Although the US over-produced automobiles by a million units, a scheduled production for 12,000 military aircraft was scaled back to 9,000 planes. "The U.S. did not produce enough arms to meet its needs in Korea," said *Time*. "It even failed to turn out enough ammunition to supply the limited force in Korea. Sample bottleneck: for lack of a small electronic part, engines could not be delivered to North American Aviation and powerless Sabre jets had to be lined up in long rows outside the plant."[6]

During the early morning of April 17 the all clear was finally given. The Nunivak sightings were never confirmed and ended up as sensational copy in *Flying Saucer Review*. The Presque Isle sightings turned out to be three commercial airliners badly off course. The Air Defense Command was criticized by the Pentagon for undue panic and overreaction. However, General Chidlaw stood his ground, saying, "Tell 'em if the situation occurs again I'll do the same thing."[7] Someone in 1941 had discounted just such a radar report of incoming aircraft at Pearl Harbor an hour before Admiral Yamomoto's dive-bombers arrived. Chidlaw had acted wisely. Ivan Getting, the MIT-trained physicist and professor of electrical engineering, who was at the time head of the Radar Panel of the Research and Development Board of the Department of Defense, realized that there was a big chink in the nation's air defense mindset that needed an equally big adjustment. "The Battle of Britain had been won through long-term attrition of Nazi bombers. However," he cautioned, "in the atomic age, a near perfect defense against the penetration of bombers was needed."[8]

George Valley, the bright boys' point man and air defense advocate, was bleaker, saying that there "was no integrated air defense system, no effective radar system for low-flying aircraft. Airborne interceptor radars failed when looking down on low-flying bombers, and while ground radars worked well at long ranges against high-flying aircraft, they could not detect low flyers."[9] Al Donovan from the Cornell Aeronautical Laboratory worked out the calculations for Valley, reporting "that a bomber flying in over the north pole region at high altitude could always detect the ground radar before the radar detected it; it could thereupon descend under the radar beam and continue undetected at low altitude."[10] Air defense in the United States was light years from Getting's warning for "near perfection." Even a single bomber getting through to a single US city would have been catastrophic.

In 1952, enemy bombers, most likely Soviet bombers, didn't carry TNT as did the Japanese at Pearl Harbor, rather, they carried 400-lb atomic bombs capable of wiping out a city with one drop, even if the bomb was quite off target. The immensely popular TV idol of the time, *Tom Corbett: Space Cadet*, showed America how to protect itself: computers, radar, and aviation technology all wrapped together. Corbett and his space goggles would have felt perfectly comfortable amid the bright boys' banks of glowing electron tubes, light pens, and display screens. General Chidlaw told Air Force chief of staff General Hoyt Vandenberg that he hoped the alert had made "more of our top Air Force people . . . aware of the very thin margin of evidence on which we too frequently must base our decisions."[11]

The alert was a false alarm but a sobering one. Both Vandenberg and LeMay were ardent first-strike advocates, believing it necessary to first bomb the enemy before they were able to get airborne. Vandenberg was also staunchly anti-air defense as well, saying before Congress that, "our greatest defensive and offensive weapon is our strategic force plus that part of our tactical force that is based within striking range of the airdromes that would be used by the Soviets."[12] He was unconvinced that a system could be built that could monitor and detect—and thereby protect—the skies over America, brushing off electronic vigilance saying, "the hope has appeared in some quarters that the vastness of the atmosphere can in a miraculous way be sealed off with an automatic defense based upon the wizardry of electronics."[13] For LeMay's part, in 1950 he had already drawn up a targeting map of the USSR with 30 cities marked off for one each of the existing 50 bombs in the US nuclear stockpile. Nunivak, however, had caught them both flatfooted in America, far from Soviet territory.

Had the Nunivak and Presque Isle sightings been real, the irony would have been compounded by the fact that Soviet bombers, TU-4s, were nearly bolt-for-bolt clones of the US B-29 Superfortress, reverse engineered by the Russians from three American planes interred in Vladivostock during the war.[14] And the recipe for the atomic bombs carried by these Soviet "Superfortreski" came in the main courtesy of the treachery of Klaus Fuchs. In short, had the Nunivak/ Presque Isle sightings produced a bona fide sneak attack, the United States would have been victimized by its very own technology. The hawkish LeMay and Vandenberg would have witnessed an exact ditto of the *Enola Gay's* ride into Hiroshima, but this time visited upon San Francisco, Chicago, or New York. The only piece of US deterrence to which the Soviets did not have the answer was purring away in the Barta Building: real-time command and control.

In 1952, General Vandenberg was well aware of the bright boys and their successes; he had bankrolled their work since January of 1950.[15] On August 29, 1949, years ahead of schedule, the Soviets set off a 20-kiloton atomic bomb. President Truman announced it to the nation in September. The nation went wild as a result, and panic set in at the Pentagon. As George Valley reflected later, "Cocksure and arrogant, most of us were fooled."[16] In the fall of 1949, vice chief of staff of the US Air Force, General Muir Fairchild, met with George Valley, and the meeting led to Vandenberg springing for some upfront experimental cash. Valley, cash in hand and looking for a computer for his air defense tests, accidentally stumbled onto the bright boys and their marvelous digital leviathan. By the time the national alert sounded on April 17, 1952, the Air Force and General Vandenberg were in deep with the bright boys, having gone so far as to finance a building for them as their own permanent research complex in Lexington, Massachusetts, naming it Lincoln Laboratory, and rushing it to completion in 1953.[17] By 1952, Vandenberg's saber rattling was no longer just hawkish braggadocio; it had a pragmatic purpose pushed along by desperate times: keep the Soviets distracted.

Perspicacity on the air defense of the United States was late in coming to the Air Force, although great concern over it had been in place immediately following the end of World War II. As early as September of 1945, General Alden Crawford had circulated a memo titled, "Necessity for an Army Air Forces Air Defense Program,"[18] in which he plainly set out the case for defending the skies over America. Its proposals, however, were fated to meet head on with the Truman Administration's cost cuttings for demobilization as well as the Air Force's own reckonings that air defense was unnecessary. Unnecessary, that is, until 1952, which is the date the Air Force figured was the earliest the Soviets would be capable of launching an air attack against the United States.

By 1947, the Air Staff had established its priorities, amongst which "Air defense missiles and detection and warning systems occupied third place,"[19] far behind the much more anticipated first and second priorities of air-to-air and air-to-surface missiles needed to effectively deliver a nuclear first strike to the Soviet Union. And for the foreseeable future that Air Force position on the matter seemed firm. Air alerts of March, 1948 and December, 1950,[20] although not as pervasive as the 1952 scare, caused major concern and opened government coffers to defense funding, yet could not budge priorities one and two. It seemed as if every two years a close-call air alert streaked into view, unnerving the military a little more each time. Being so vulnerable and feeling it, might well have provoked a dangerous reaction.

Physicist Guy Stever was a first-hand witness to it all. His book, *In War and Peace: My Life in Science and Technology*, illuminates much of the backdrop of events surrounding the travails of air defense. Stever began his affiliation with the Air Force directly after the landings at Normandy, during which, as a special scientific agent on loan from MIT's Radiation Lab (developer of Allied microwave radar), he followed close behind the troops in a hunt for advanced technology left behind by the retreating Germans. After the war and back at MIT, he was indispensible to military programs in aerodynamics, aviation, guided missiles, and intercontinental ballistic missiles (ICBMs). From its founding to 1968, he was a member of the Air Force Scientific Advisory Board (SAB), and he was chief scientist for the Air Force from 1955 to 1956. If anyone had a view as to what the bright boys were up against, it was Stever. He, like George Valley and the Air Force, was stunned by the Soviet's atomic bomb test of 1949, lamenting that the explosion "also exploded comforting illusions: that our nuclear monopoly made us impregnable, that we could safely and sharply cut our spending on defense, that we were technically superior."[21] Years were lost before General Muir Fairchild, in November of 1949, called an emergency session of the

SAB to reorient priorities. As Stever saw it, "Our defenses were bad. The Soviets had the bomb, and the United States essentially had no effective defenses against bombers carrying nuclear weapons."[22]

Men like Vandenberg and LeMay—two of the top dogs in the world's best and most fearsome air force—persisted in their first-strike beliefs. Products of their time, they spent nearly four years of World War II bringing nonstop aggression to difficult enemies, much of it in desperate hours, some of it without much hope at all. After the war, theirs was the first group of battle-hardened senior officers to command the new Air Force. They had survived and won the air over Europe and the Pacific and knew well the price in countless comrades lost. They knew before anyone that the deadly voyage of the *Enola Gay* would change everything forever. They, before anyone else, were the first to view the aerial photography of Hiroshima and Nagasaki; they remembered the pin-drop quiet of the operation's room as they closed their eyes and gasped, My God! Their eyes reopened to a different world, one where entire cities and whole populations could be obliterated in an instant; vaporized beneath the haunting dark plume frozen in the photographs before them. Their resolve was never to have that horror visited upon their own. Not on their watch! Even if it took a first strike against an adversary to ensure it.

Contrary to Hollywood-heightened caricatures of trigger-happy megalomania, the Air Force high command has always been well stocked with levelheaded, oftentimes brilliant, but very aggressive personalities. It's the bloodline. This newest of military services charged with pioneering the newest frontier, air power, has attracted many of America's best with its allure of glamour and risk in its "footless halls of space."[23]

Since its inception, the action-oriented, maneuver-and-attack mindset has been integral to its success. Major General George Squier, who in 1917 headed the Army Signal Corps Aviation Section, was an early Air Force archetype. General Squier held a doctorate in physics from Johns Hopkins University, held numerous patents on electrical communication devices, and had a yen for barrel rolls and aerial adventure. Robert Millikan, the Nobel laureate in physics who worked with Squier on World War I's National Research Council, described him as a "strange character . . . who had one great quality much needed at the time, namely, a willingness to assume responsibility and go ahead."[24] Much the same can be said about airmen like Billy Mitchell, Hap Arnold, Tooey Spaatz, Jimmy Doolittle, and Bernard Schriever.

What the Air Force high command did have by 1948 was license to attack. It came in the form of National Security Council Memorandum 30

(NSC-30), which established the policy of "prompt use," a euphemism for the doctrine of preemptive strike. It was, as Paul Edwards describes it in *The Closed World*, a plan for "an all-out nuclear attack against the USSR in any situation where it appeared the USSR might be about to launch a strike of its own."[25] Responsibility doesn't get much weightier than that. But it was their watch; they took their duty seriously; and they were the right bunch of guys to get the job done. It was much to their credit that in the intervening years no one ever exercised that grizzly option, not even during the trigger-twitching uncertainty of Nunivak.

Well before Nunivak, no part of the Soviet arsenal of airborne destruction was a surprise to US intelligence. During the August 3, 1947 Aviation Day over Tushino Airport in Moscow, the Soviets had flown the cloned B-29s for all the world to see.[26] After ground testing their first atomic bomb in 1949, the Soviets successfully test-dropped another from a TU-4 over Kazakhstan. Given the designation "Bull" by NATO, the TU-4 Bull entered the Soviet Long-Range Aviation Service (the *Dal'naya Aviatsiya* or *DA*) in 1949, and by 1952, the number of planes that were operational was 850.

Fitted for midair refueling, the TU-4 had the capability of striking the United States, Europe, North Africa, the Near East, and Japan.[27] Some of the TU-4s were even capable of pulling along behind them on tethers MiG15s as potential fighter escorts. The Soviet Long-Range Aviation Service was also well aware of the limitations of US radar. Soviet TU-4s could and would have easily avoided detection at Nunivak or Presque Isle. Flying low, the TU-4s could have slipped into the United States in any numbers just about anywhere they pleased. To Getting and his mates at the Pentagon, it seemed like the jig was up: "The Korean War was about to explode into a third world war with the Chinese coming in, and the Soviets were about to bomb us with their long-range bombers. We were no longer a safe island, isolated from the rest of the world."[28] Surely, a repeat of the Nunivak event must somehow be prevented: the incident was expensive, dangerous, a communications blunder, untrustworthy, and it was just too damn uncertain. Air defense, what there was of it, could not continue being, in the lingo of Tom Corbett, "as unstable as a heavy isotope."

What if the crew on Nunivak had used the bright boys' technology? How would the events of April 16th and 17th of 1952 have played out differently? And just exactly what was it that made this Tom Corbett stuff in the Barta Building so very special?

The exterior of the bright boys' system seemed ordinary enough with its radar dishes and antennae. There were large, long-range radar that sent microwaves out to the horizon looking for something to bounce off of and return a signal. Two other radar types helped out: gap-filler radar to fill in open gaps between the long-range radar, and height-finder radar, which determined the altitude of incoming aircraft. Between the three radars, both high-flying and low-flying aircraft could be revealed and tracked in real time, including their latitude, longitude, speed, and bearing.

The remarkable part is what happened after detection took place. The radar signal data would be sent by telephone lines and downloaded at the Nunivak command center via modem into a gigantic, room-sized digital computer called Whirlwind. This communications process, heretofore never attempted—from radar to telephone lines and then downloaded into a computer— represented the first-ever digital network.

Whirlwind, in turn, would store real-time signal data on all of the approaching aircraft. Whirlwind was the bright boys' real gem: the first real-time digital computer. On large circular, display screens Whirlwind would show the Nunivak crew all of the aircraft in their sector as spots of light on monochrome screens. If a Nunivak crewmember then pressed a special light gun onto any of the screen spots, Whirlwind would instantly begin tracking the screen object, spitting out a continuous stream of flight-approach information. Whirlwind could also instantly plot intercept courses, sending them directly to Tactical Air Command to scramble jet interceptors to the targets. Eventually, in 1959, the system would also have the ability to direct Air Force BOMARC surface-to-air missiles to targets, and when tied into ARADCOM's Missile Master System, direct Nike-Hercules surface-to-air missiles as well.[29]

Collectively, the bright boys' digital computer, with its detection and communications systems, would come to be known as SAGE for Semi-Automatic Ground Environment. From its 24 Direction Centers spread out over the continent, SAGE would keep everyone on the same page. From Nunivak to Elmendorf to Ent as well as across country to Presque Isle, everyone would be aware of what was happening as it was happening—and be able to mount a timely and meaningful counterattack. The system's modems, display screens, light guns, and computer keyboards were first-ever uses of the technology, yet represented barely a portion of the electronic trove hidden behind the walls of the Barta Building.

The critical advantage possessed by any TU-4s edging into North American airspace would not be in military armaments; rather, it would be in the form of an information advantage. The Soviet Long-Range Aviation Service knew to fly high and then to fly low to avoid US radar; information gleaned from B-29s gave them a delivery system capable of reaching American cities; and the knowledge to build the nuclear terror in their bomb bays came courtesy of espionage. It was an advantage that could only be offset by the United States possessing better information than the Soviets.

Following Nunivak, air defense took a dramatic turn toward information preparedness. A string of radar stations across Canada called the Pinetree Line was begun. Topping that was a recommendation made by scientists from the Summer Study Group meeting at MIT in 1952.[30] They called for the immediate construction of the now-famous Distant Early Warning—or DEW Line—of radars across northern Canada. The DEW Line plan was for an integrated chain of 63 radar and communications stations, all within the Artic Circle stretching over 3,000 miles from the coast of Alaska to Baffin Island. Begun in December of 1954 and completed in July of 1957, the DEW Line involved over 25,000 workers and 460,000 tons of material.[31] The end result was a string of sensors sweeping the skies for any hint of encroaching Soviet aircraft.

The resulting information increase, however, was astronomical: a glut that once captured still needed to be delivered southward to the US Air Force for filtering, analysis, and decision making. If a single station like Nunivak had an almost impossible time communicating via Teletype and telephones, what then of 63 stations doing it simultaneously? Paul Bracken, an analyst of defense technology and innovation, figures that such improved detection offered the military an increase in tactical warning time of an hour or more.[32] And for someone like General Chidlaw, the luxury of an hour of decision time would be very much welcome. To make it all work took implementing some advanced technology and removing many of the human operators from the process—basically automating most everything from radar sensing to communication to decision making. The only available technology to give the DEW Line such a true information advantage was SAGE—powered by its digital engine, the Whirlwind computer.

First begun as an analog computer project in 1944[33] and recast as a digital computer in October 1945, Whirlwind by 1949 was on the ropes and about to be cancelled as an expensive failure. Luckily it wasn't. In the nick of time it was discovered by the Air Force, who ponied up the cash to make Whirlwind a reality. It's first-ever system's test in October of 1950 was a complete success and total vindication of the bright boys' dogged belief in themselves and their invention. However, between 1948 and January of 1950, the bright boys stood alone in keeping their machine from the scrap heap. Two of the twentysomethings, Jay Forrester and Robert Everett, led the way, holding the fragile project together until a chance encounter with

MIT physicist George Valley, who held the Air Force's checkbook. The adventure that followed is one of the greatest in the annals of engineering history.

Ironically, the reason why the bright boys' whiz-bang technology wasn't spotting Soviet bombers sooner than Nunivak was because of an Air Force bomber, one that 15 years earlier had landed at Langley Field, Virginia.

Paradigm Lost

On March 4, 1937,[34] airmen at Langley Field heard a deep drone resonating overhead. At 5,000 feet in a heavy, overcast Virginia sky, a huge four-engine aircraft circled and then descended toward the runway. The airmen below were bedazzled by the giant silver/aluminum bird swooping down at them: such a thing of mechanical genius, beauty, and breathtaking military prowess.

The noise from its four 1,000-horsepower Wright Cyclone engines grew ever louder the closer it got to the ground. General Frank Andrews, commander of the Army's General Headquarters Air Force, and his fellow bomber zealots, watched with prideful expectation as their brand new big gun made its approach. The Second Bombardment Group had waited a long time for this moment, and they were more than ready and willing to take on the challenges that lay ahead. The Army Air Corps now had its mitts on the biggest, fastest, highest flying, longest range, and best armed strategic bomber in the world: the Boeing YB-17, soon after redesignated as simply the B-17.

At a top speed of 256 mph, it was faster than most Army Air Corps pursuit aircraft of the time, most of which were barely a fraction of this heavyweight's 34,000 pounds. It bristled with five machine guns (nose turret, blisters on the top and bottom, and one each on either side of the fuselage at its waist), and could carry five tons of bombs at 30,000 feet to any enemy up to 1,300 miles distant and then fly home again. At the controls was Major Barney Giles, who had picked up the big bird fresh off Boeing's Seattle assembly line.[35] Dubbed the "Flying Fortress" by a Seattle newspaper reporter, the gleaming giant, 68 feet long with a wingspan of 103 feet, roared onto the Langley runway. This, Andrews and his fellow airmen must have thought, is air power.

They were so confident in the bomber's enormous potential that the Chief of the Air Corps, General Oscar Westover, would later claim that, "no known agency can frustrate the accomplishment of a bombardment mission."[36] Much the same was claimed for cavalry equipped with Winchester rifles by General Custer. But just maybe this giant was different. It was hard to imagine that a hundred or so B-17s all flying together in formation could be stopped by anything of this world. With its four engines ramming it through the air, machine guns blazing, and bomb bays dropping steel from crazily high altitudes, the big bird, indeed, seemed awesomely invincible. And nothing would soon dissuade the bomber zealots from that conviction.

Years of experience pitting pursuit aircraft against bombers were tossed aside. So too was common sense. Airmen were on a giddy, high flight of zealotry that would linger on well past Nunivak. Bomber invincibility was all about having the baddest and best big stick on the block. This was vintage bare-knuckled one-upsmanship just begging to be taught a stiff lesson. During World War II, the Army Air Corps compounded the braggadocio further by adding to the mix what was called in the war plans HADPB, an acronym standing for High-Altitude Daylight Precision Bombing.[37] HADPB was all about flying over the enemy's territory in broad daylight, maybe even high enough to avoid deadly flak, and then dropping bomb loads precisely on the desired targets. HADPB was a major pipedream.

Practice bombing—in the relative calm of the United States without the abrupt discomfort of flak or 20 mm cannon shot ripping through airframes—was less than promising; it was a joke. Such was the case with the famous Pee Dee River bombing of 1927.[38] A reinforced concrete bridge spanning the river was donated to the Air Corps by North Carolina. A day of bombing using 300- to 600-pound bombs from 6,000 and 8,000 feet did scant damage to the span. Bombing was a tough job: bombers are in motion, the Earth is turning, and dumb bombs must angle between the two to hit a target. Even after ten years there was little improvement. Lt. Colonel Harvey Burwell experimented at Muroc Dry Lake, California in 1938.[39] Accuracy again failed to materialize. Bomber fanatics were in for a rude awakening when some finally realized that bombing should never ever be used in the same company with terms like precision, high altitude, and daylight.

HADPB using B-17s without fighter escorts deep over the German homeland would crash headlong into the Kammhuber Line of flak towers, blistering antiaircraft fire, radars, and skies buzzing with enemy bogies. It was a bloodbath of unprecedented proportions. Thousands of bombers, each carrying a 10-man crew, flamed earthward. More American airmen would perish in the skies over Europe than in all the Marine landings in the Pacific.[40] It would take a horrendous 25% casualty rate to stop the insanity. Only six months into 1944, by June 6th, air supremacy over Europe was finally achieved, but at a cost of 2,700 bombers, 1,000 fighter aircraft, and over 18,000 casualties. In all, such losses were 50% greater than the total of aircraft downed in 1942 and 1943 combined![41] Unfortunately, even post-WWII, the dogma of strategic bombing would remain firmly entrenched as the Army Air Force's reigning paradigm for making war.

Scientific study has a corollary to such organizational behavior. *The Structure of Scientific Revolutions* by Thomas Kuhn explores what Kuhn calls *received beliefs*[42] as being most important to scientific communities in forming a fundamental foundation about what they believe. "No natural history," writes Kuhn, "can be interpreted in the absence of at least some implicit body of intertwined theoretical and methodological belief that permits selection, evaluation, and criticism."[43] This body of received beliefs he calls a *paradigm*, and he argues that it exerts such a "deep hold" on the practitioner's mind that he will strongly resist any new paradigm or theory that conflicts with it. Such was the state of classical physics until the revolution of Maxwell's equations replaced it with formal physics.

In an adaptation of this Kuhnian paradigm, one developed by the physicist Fritjof Capra and geared toward organizations in general, Capra defines a paradigm as "a constellation of concepts, values, perceptions and practices shared by a community, which forms a particular vision of reality that is the basis of the way a community organizes itself."[44] But what if the

Aerial view of Air Force Tactical School, Maxwell Field, Alabama, circa 1938. (Photo Courtesy of USAF Historical Records Department.)

paradigm is a bit queer and outdated? What then of the community organized around it? A particular vision of reality from a learned mob is what sent Giordano Bruno to the stake for the heresy of heliocentrism. Such a mob vision was the set of received beliefs held fast by the Army Air Corps. Scientific revolutions occur, adds Kuhn, when an "anomaly appears that subverts the existing tradition."[45]

Such an anomaly loomed just inside the gates of Maxwell Field, Alabama, at the Air Corps Tactical School (ACTS) But in 1937, at the height of bomber frenzy, that anomaly—nascent command and control—seemed light years from reality. Its bearers, a couple of Air Corps bright boys, Major Claire Chennault and Captain Gordon Saville, were rejected by the reigning paradigm of strategic bombing. For his part, the disgruntled Chennault resigned on disability on April 30, 1937 and went to China. He itched to be his own man somewhere and got the chance when he formed the *Flying Tigers*. Saville stayed put. "It was one of the most disgusting and disheartening periods I ever went through," he admitted years later. "Once in awhile, I was just tempted to turn in my suit and say, 'The hell with it.'"[46] The paradigm of strategic bombing steamrolled everything in its path until late in 1943, when it ground to a shattering halt in the aerial slaughter that was the sky over Europe. Although battered and bloodied, bomber zealotry staggered on for yet another decade thereafter.

Major Chennault and Captain Saville birthed a new paradigm of using information to direct aircraft in battle. Gleaning information telephoned to central command from what they called an *information net*, which was a carefully placed geographic grid of ground observers, they would assemble and plot on a map the coordinates of incoming aircraft and then vector pursuit planes to intercept them. To be sure, ground and aerial observation were not newcomers to war; each had played prominent roles in many battles long before Chennault and Saville. Their novelty was in honing observation into a coordinated picture of the entire airspace and then using that picture to tactical advantage. Chennault used the new techniques to complement his ideas on formation flying of pursuit planes in order to gain air supremacy. Saville was more defensive oriented, seeing the paradigm as the key advantage in defending an area from air attack. Both of their views, however, together with modern pursuit aircraft to do their respective jobs, would one day rule the skies. But not in the Army Air Corps of 1937. To airmen who believed that bombers were invincible in conducting and succeeding at any assigned mission, such a developed picture of the airspace was totally unnecessary. Chennault and Saville were left to dangle and their inchoate ideas on command and control to atrophy.

Chennault's only outlets for propounding his theories and observations were scads of letters to anyone who cared to read them, reams of reports to the command staff, articles published in the venerable *Coast Artillery Journal*, and his well-overlooked textbook, *The Role of Defensive Pursuit*. Regarded as abrasive, cocky, and too outspoken by his superiors as well as most other bomber zealots, Chennault's theories fell on deaf ears—except for those of Saville. He would have to wait until his *Flying Tigers* were airborne over China or at the defense of Rangoon to test his theories in battle.

In a 1997 research paper for the Air Command and Staff College, arduously but aptly titled, "The Controversy Behind the Air Corps Tactical School's Bombardment Theory: An Analysis of the Bombardment versus Pursuit Data between 1930–1939," Major Hugh Severs examined many of Chennault's writings. One letter from 1934 to his commander, the then Lt. Colonel Henry "Hap" Arnold, is of particular significance as a futurist vision. Chennault succinctly sets out the new command and control paradigm. Severs writes that Chennault's arguments were convincing in general and especially so when concluding his letter to Arnold, saying that "To evaluate a defense's effectiveness against a bomber force in a particular scenario, a systems approach is required that considers the entire defense system which includes aircraft, antiaircraft artillery, air and ground information nets [air reconnaissance and ground observers], tactics, organization, and other factors pertinent to the exercise."[47] Chennault's "systems approach" was a veritable recipe for command and control, and a key, early-on insight to making air power truly successful. With Chennault's departure, Saville was left virtually alone to carry the message forward.

In 1937, with Chennault headed to China to form his *Flying Tigers*, two events would crystallize Saville's thinking. On April 27, 1937,[48] the German *Luftwaffe*, masquerading under the name of the Condor Legion during the Spanish Civil War, bombed into oblivion the Basque city of Guernica. The Condor Legion, for all its modern equipment, tactics, and intricate plans for masses of Heinkel He-111 fighter-bombers and Messerschmitt Bf109s, failed to use ground-to-air radio to direct any of the bombing during the three-hour raid. If the Germans had not thought to use radio to direct attacks, then Saville felt confident that his newfound tactic might one day prove to be of critical advantage. A second and more important event for Saville took place one month later on the night of May 26, 1937[49] at Fort Monmouth, New Jersey. Civilian engineer John Marchetti and his Army Signal Corps SCR-268 radar unit were successful at detecting a B-10 bomber 12 miles out to sea. For Saville, radar's immediate impact was that it pushed the frontlines forward. By 10 miles or 100 miles, radar enabled defenses to begin gathering information on the enemy much earlier. And in response to intruders, pursuit aircraft could be vectored on intercept courses with reasonable accuracy. Interestingly, one of the observers at Fort Monmouth was none other than Hap Arnold, who in September of 1938 would become commander of the Air Corps, and four years later in World

General Gordon Saville, U.S. Air Force, circa 1949. (Photo courtesy of USAF Historical Records Department.)

War II, would become Commanding General, Army Air Force. Although Fort Monmouth made Arnold a believer, the juggernaut of history was ahead of him and moving away too fast to overtake bomber invincibility.

The Monmouth experiment fulfilled Saville's tactical scheme. A centralized command would take incoming radar sightings of attacking bombers and then relay those coordinates via radio to synchronize fighter aircraft and vector them for attack. With that brilliant coupling came the birth of command and control. As Kenneth Schaffel in *The Emerging Shield* succinctly puts it, ground control went from "air liaison" to "air command."[50] The bright boys in the Barta Building would roll into the mix their Whirlwind digital computer, jumping up command and control to that of "real time," where radars by the hundreds could track aircraft sightings instantaneously by the thousands.

Saville's ideas on a systems approach were not immediately embraced or even popular. Rumor had it that some pursuit commanders were none too pleased with ground control invading their cockpits. Not until, that is, the British in 1940 used radar and radios in the Battle of Britain to defeat the *Luftwaffe*. Command and control is what turned the *defense* of Britain into the Battle of Britain. The British had used command and control's defensive capability as a weapon. And their ammunition was information. Suddenly in 1940, Gordon Saville was a genius. After the war, both Saville, then as a general, and Marchetti, then regarded as the Father of American Radar, would reconvene their skills in the Barta Building. Until then, the new paradigm was lost.

Enter Arnold

As he sat before President Franklin Roosevelt on September 28, 1938,[51] General Henry "Hap" Arnold had commanded the Army Air Corps for all of

General Henry "Hap" Arnold, U.S. Army Air Force, 1945. (Photo courtesy of Robert and Kathleen Arnold Collection.)

seven days. The very next day, September 29th, he would officially take leadership of 1,650 officers and 16,000 enlisted men. The previous Tuesday, September 27th, Arnold's boss, General Oscar Westover, was laid to rest at Washington's Arlington Cemetery. Westover was killed attempting to land his A-17AS attack plane at the Burbank, California airport on the 21st of the month. At the meeting, the president ordered Arnold to build 10,000 new military aircraft. For Arnold, who had barely 1,700 total aircraft under his command from an American aircraft industry that had never built more than 2,000 in any one year, 10,000 was a mighty tall order.[52]

Life as an airman was suddenly moving quite fast for Arnold. Having entered West Point at the age of 17 in 1903, the intervening 35 years of his military career had spun themselves out at a much more leisurely pace than the previous seven days. He was 52 years old and totally white-haired by 48 after several heart attacks. He had endured the humiliation of exile to Ft. Riley, Kansas for testifying on behalf of his friend Billy Mitchell. Since 1936, he had bided his time and held his tongue about an independent air force from his boss, Westover, a loyal Army man. Arnold now found himself tossed headlong into the biggest job of his life: getting his country's air force ready for a world war. The six-foot Pennsylvanian was a no-nonsense leader, highly experienced at command, a first-rate airman with a visionary's bent for aeronautical technology, and an incessant advocate of air power and an unfettered air force. In the months and years ahead he would need all of it and more.

While his meeting was in session, six time zones away in Munich, Germany, European leaders were also meeting, attempting to appease the vainglorious Adolph Hitler and save Czechoslovakia from losing a chunk of itself to the Nazi fuehrer. In 1938, Hitler was on a roll. By that September, he had remilitarized the Rhineland, had annexed his native Austria, had helped General Franco regain power in Spain, had flaunted the Treaty of Versailles by building a powerful new *Wehrmacht,* and now demanded the handover of the *Sudetenland* from the Czechs. The British Prime Minister,

Neville Chamberlain, flew to Munich looking to secure "peace in our time" for which Hitler would gladly reciprocate by signing a "friendship" note for him to take home to wave at a relieved British public.

The German general staff knew otherwise. Since April, Hitler had directed them to begin planning *Case Green*—the invasion of Czechoslovakia, announcing that, "It is my unalterable decision to smash Czechoslovakia in the near future by military action! We will tackle the Czech situation, then we will tackle the situation in the West."[53] At Arnold's meeting, Roosevelt already knew that on August 17, 1938, Count Ewald von Kleist-Schmenzin had made a secret visit to London, arranged by the British foreign correspondent Ian Colvin, to convince Chamberlain of the German general staff's intention to overthrow Hitler. Czechoslovakia was an ally in an entente with Britain, France, Poland, and Russia, and the German military knew that a fight over Czechoslovakia would drag Germany into a war in which it would single-handedly have to fight five enemies at once. The meeting with Chamberlain never took place. The best von Kleist-Schmenzin could do was to get an interview with the leader of the Tory opposition in parliament, Winston Churchill.

However, the die was cast. The day after Arnold's meeting, Thursday, September 29th at 1:30 a.m., Hitler, Chamberlain, Mussolini, and French President Edouard Daladier signed the Munich Agreement. In a meeting to which Czechoslovakia was not invited and did not attend, where an agreement was reached that Czechoslovakia neither helped to author nor was permitted a vote upon, Britain and France officially agreed to surrender the Sudetenland to Germany. The Czech president Benes was urged to sign it immediately, and he did. Hitler got the Sudetenland with its population of two million ethnic Germans, and six months later *Case Green* was sprung on what remained of the Czech homeland. Securing "peace in our time" would now fall to the efforts of others like General Arnold and his airmen. Command and control, already in the backseat of Air Corps' affairs, was now headed for the rumble seat. HADPB and B-17s were about to take over the wheel.

Polar opposite to the American lack of interest in command and control were Britain's great interest, sense of urgency, dedication, and real success in getting their newly hatched command and control brainchild to work. Gordon Saville would have felt very much at home in the England of 1938.

By 1938, the British had built a network of integrated radars snaking along the English Channel from Ventnor near Portsmouth in the south, to

Netherbutton in the Orkneys to the north.[54] Called the Chain Home system, it could detect the bearing and distance of incoming aircraft, in some cases for up to 100 miles away. The radar waves or wavelengths at the time were long and broad, and as such, were not accurate enough to determine the number of aircraft in a formation. And with antennas in fixed positions seaward, not much coverage over the interior was possible. Once past the Chain Home electronic barrier, intruders had to be tracked by ground observers, known as the Royal Observer Corps. This floodlight-type wavelength approach sprung from the necessity of getting radar up and working quickly. The leaders of the radar research, Robert Watson-Watt and Arnold Wilkins, would have preferred to search for a narrower beam able to discern individual aircraft.

Air Marshall Sir Hugh Dowding, Commander-in-Chief of the Royal Air Force (RAF), Fighter Command came up with a brilliant plan to tap into the Chain Home's sightings. He correctly realized that radar was useless without a command and control system to receive and act upon the acquired information. He created what the RAF called a "filtering" system[55] Radar and Royal Observer Corps sightings were telephoned to the central command center at Bawdsey Manor in Felixstowe, Suffolk, which Dowding called a filter room because it *filtered out* information from many different sources. Like wartime Hollywood fare right out of a Ronald Coleman movie, Greg Goebel's scene description of the RAF command center in "Radar and the Battle of Britain"[56] nicely conjures the feel of the filter room:

> In the center of the filter room was a huge table, with a map of the British Isles and neighboring coastal regions of Europe. Staffers moved markers across the map with tee sticks to track the movements of the intruders. A controller observed the movements and determined which fighter squadrons to scramble. Alerts were sent to the four Fighter Command groups by telephone or Teletype. The system was clumsy by modern standards, but a clever and impressive feat of organization for the time. As the *Luftwaffe* pounded Britain, the filter room became the central operations center for air defense. Reports of intruder formations tracked by radar and Royal Observer Corps networks were relayed into the filter room on dedicated military phone lines. Officers logged the reports and assigned a 'battle number' to each formation.

The British had thought out the implications of networked radar technology and masterfully put it to best use. And as with many other scientific breakthroughs, there were unintended consequences aplenty—rather pleasant ones. Eddie Bowen is just such a case. His work on the Chain Home radar intersected with his interest in airborne radar, which he perfected and named Airborne Interception or AI, which, in turn, intersected with yet another of his interests, ocean surveillance, leading him off again into air-to-surface radar, called ASV.[57]

Later that year the RAF command and control center was moved to the London suburbs at Bently Priory. And as it settled in, yet another benefit popped into being almost immediately: Operations Research (the British call it "Operational Research"). J. E. Beasley, Professor of Operational Research at Brunel University, West London, defines it as "scientific analysis of the operational use of military resources."[58] Basically, after the physicists and engineers set up the technology and got it running, the OR team entered, analyzed the system's performance, and then modeled it to optimize its use. Sir Charles Goodeve has a practical name for it: "quantitative commonsense."[59] If, for example, most enemy aircraft came from a single direction like Calais, then how should the entire system best utilize its resources to react to that situation? A. P. Rowe, chief superintendent of all radar research projects, instituted operational research—what is today nearly identical to systems analysis—following a poor performance of a Chain Home system test. Air Marshall Sir Hugh Dowding was so impressed he permanently attached them to his command and designated the groups as the Operational Research Section.[60]

Operations Research was a byproduct of command and control and would grow in importance as the war progressed. The Section's analysis work was highly adaptable, covering needs like modeling depth charge lethality in hunting U-boats, to maintenance cycles for aircraft, to the necessity of low-level bombsights. Rowe also recognized early on that strong relations between the military and Chain Home scientists were critical in building trust, providing mutual feedback, and for follow-up ideas. He began informal meetings between the two, which he called *Sunday Soviets* (from the Russian word for *council*), holding them weekly on Sundays at the Grosvenor Hotel.[61]

Eventually, the British would pioneer command and control's first killer breakthrough: wide-beam radar. This was followed by an even more remarkable discovery by two University of Birmingham engineers, Harry Booth and John Randall: a device called the cavity magnetron that produced microwaves for highly accurate, narrow-beam radars. Guy Stever relates that the palm-sized device "emitted highly focused microwaves at great power . . . roughly a thousand times the output of the best US tube on the same wavelength."[62] The days of radar's wide waves were numbered. The cavity magnetron "meant greater accuracy, better resolution, less clutter from extraneous echoes, and less atmospheric interference."[63] To perfect it into a wide range of microwave devices, Sir Henry Tizard secretly brought the magnetron to the United States. MIT's Radiation Laboratory (Rad Lab) took it on and succeeded at perfecting microwave radar for the war effort as well as MIT's postwar renown as world leader in the field.

Stever, with "a freshly minted CalTech PhD in physics," became badge #149 at the Rad Lab. On New Year's Day 1943, Rad Lab director Lee Dubridge sent the 25-year-old to the OSRD's London office to liaison with British scientists for the transfer of more of their radar expertise. He and David Langmuir were assigned to work with the British armed services, most notably the Royal Air Force as it began to implement "cavity magnetron technology for ground radars, airborne interception, bombing, and navigation systems."[64] Frequently, Stever trained over to Great Malvern on the Welsh border, where British radar R&D was centered. To get there, he would take the London–Great Malvern route, riding the Paddington train on its long journey west through Cotswald to Worcester. Much of the journey was spent at chess with his frequent partner and fellow physicist, George Valley, who was at that time about seven years distant from discovering the bright boys and their Whirlwind computer.

"The American radar program," writes Stever, "had at the time nothing like the tight coupling by the British of research and development to military needs and operations. British laboratories were in close contact with their frontline airfields, naval ports, and army units."[65] The US Joint Chiefs of Staff responded by establishing its own Committee on Radar Research and Development headed by MIT's president, Karl Compton.[66] The free flow of British command and control expertise—both in terms of technology and relationship building between science and the military— began streaming across the Atlantic. Saville got himself to England twice before Pearl Harbor and was impressed by the Chain Home in full operation, remarking that the British were "far, far superior to what our people were doing."[67] In 1950, command and control would meet up with its next and maybe greatest killer breakthrough: the digital computer. Much of the Chain Home/Rad Lab development ethos, right down to clones of Sunday Soviets, would wiggle in through the old bricks of the Barta Building and also remain firmly in the thoughts of Saville.

The primordial ooze from which the bright boys would later bring forth real-time digital computing and real-time digital networking—and mate them to the needs of real-time command and control—was so close yet so very far away in wartime England. In 1935, at Cambridge University, Alan Turing invented the principle of the modern computer. The previous year, Thomas Flowers had built electronic digital equipment incorporating 4,000 electron tubes to control telephone connections at the British Post Office Research Station at Dollis Hills. Then in 1938, Flowers followed that up with a companion high-speed, digital data-processing system. It would be these high-speed digital techniques using electron tubes that would make the modern computer possible.

Alan Turing and Gordon Welchman in early 1940 (Welchman would find his way to the Barta Building after the war) were using electromechanical machines (built on earlier Polish models) to decipher 39,000 intercepted German radio communications per month. Cambridge University's Max Newman in 1942 proposed automating the decryption process with a high-speed counting device, which would turn out to be the Heath-Robinson computer. Flowers would augment the Heath-Robinson device with the British-pioneered, thyratron electron tubes to birth the famous Colossus, the first fully functioning digital calculator. Many would later call the Colossus the first digital "computer." It was not. Maurice Wilkes, the postwar, British computer pioneer, points out that since its operations "were logical and not arithmetical," the Colossus is more appropriately defined as "a special-purpose digital signal processor."[68]

Closer approximations of modern digital computing machines up to 1941 were more fully realized in Berlin with Konrad Zuse's Z3 or at Iowa State University with John Atanasoff and Clifford Berry's ABC. However, the twin developments of radar and digital data processing, so very critical to the future of real-time command and control, were technologies first developed in a not-so-Merrie England as it girded for battle. Much of that early British pioneering would cascade westward via OSRD's London office, primarily coming to rest in Cambridge, Massachusetts, and into the incredibly fertile awareness of 2,000 researchers at MIT's Rad Lab (in residence were 14 future Nobel laureates). The Rad Lab itself was the product of Vannevar Bush's OSRD foresight and the $450,000 he put up to get it all going.[69] Laboring along with Guy Stever's badge #149 were badge numbers assigned to the bright boys' future brain trust of Jay Forrester and Robert Everett at the nearby MIT Servomechanisms Laboratory.

A great disappearing act took place in postwar England as both microwave radar and computing went into eclipse. It would have been beneficial to both technology and the researcher if something like the strong London to Rad Lab cross-fertilization had continued in earnest well after the war. It was not to be. In Britain there was a brief spurt of luminance in computer building between 1946 and 1953 that produced the Manchester Baby, EDSAC, Pilot ACE, and LEO computers, but that was about it. The Lyons Tea Company (think Starbucks coffee shops) used its LEO, the first business computer, to do the first-ever computerized payroll. The yearning and talent were obviously there and with a little bit of assistance a few Barta Buildings might have sprung up, some with the staying power that every new craft needs to carry itself over to more prosperous times.

The government secrecy imposed on Colossus until 1976 by the Official Secrets Act surely did not help the spread of the new technology. Worse still was the near absence of financial support from the National Research and Development Corporation (NRDC), a government scientific funding office begun by then Trade Minister Harold Wilson in 1949 (Wilson later became Prime Minister in the 1960s). The NRDC was supposed to do much, but did very little investment. In its first ten years, the NRDC distributed cash and loans to the new computer industry that amounted to a paltry $4 million, while the US military over the same period ponied up over $1 billion for American computer development.[70] John Hendry's *Innovating for Failure: Government Policy and the Early British Computer Industry* is a scathing evaluation of the NRDC's inept bungling and "help to no advantage." The pity is that computer technology went without much of that quirky, eccentric brilliance that the British seem always to muster up to solve problems in, or to illuminate, a new technology or science. Lamentable was the talent of Turing, Flowers, Newman, Kilburn, Wilkes, Williams, and the rest of the richly talented British bright boys left to dangle at a time when they might well have pushed out a couple of potentially great computer companies.

Something had to give in a war-ravaged country trying desperately to aright its economy, so a homegrown computer industry became a big casualty. And Albion most definitely was war ravaged. Five years of aerial bombardment ended on March 29, 1945, during which "the enemy had dropped a grand total of 70,995 tons of high explosives and uncountable thousands of incendiaries, killing 60,595 civilians and seriously injuring another 86,182, destroying 222,000 homes and damaging nearly five million more."[71] E. R. Chamberlin's *Life in Wartime Britain* grimly describes the 76 continuous nights of the London Blitz, where in "central London only one house in ten escaped damage."[72]

The United States had no such burden of injury; high-tempo research in microwave radar and digital computing was free to flourish. The technology transfer to the Rad Lab that dispersed itself from Cambridge to Washington catapulted the United States into a leadership position that it never relinquished.

Postwar, the Air Force would have a Strategic Air Command and a Tactical Air Command, but its Air Defense Command would be disbanded. Air defense was deemed unnecessary.[73] Ground observers were sent home, radar networks mothballed, and the Air Force returned to being all about bombers. MIT's Radiation Laboratory was shut down in 1945, with its 2,000 professional staff heading off to normal lives once again. In 1947, the United States Air Force officially came into being and was, without question, the most powerful air force ever created. However, two years later, in 1949, with the Soviet Union's first mushroom cloud billowing upward, a desperate search began for something other than raw air power. America's quest for its own version of the Chain Home would begin in earnest. The first steps on that journey were traced to the old laundry building on Massachusetts Avenue by Guy Stever's friend and chess partner, George Valley.

Chapter Three
Real-Time 'Digerati'

*Formula for breakthroughs in research: Take
young researchers, put them together in virtual seclusion, give them an unprecedented
degree of freedom and turn up the pressure by
fostering competitiveness.*

—James D. Watson, Co-discoverer of DNA

On June 6, 1946, 28-year-old Ted Williams hit the longest
homerun—502 feet—in the history of Fenway Park. Seat #21,
where it hit in the Fenway bleachers, was painted red
to memorialize the feat and has remained that way ever
since. On the same day three miles away, 28-year-old Jay
Forrester, 25-year-old Robert Everett, and their crew of
budding digerati were unknowingly about to hit out a long
one for digital computing: they were getting ready to invent Information Technology and to create the world's first
Information Technology company.

A near-impossible concept for anyone to fathom in 1946
would have been that of Information Technology. Information
itself was only of marginal interest, and then mostly as
chunks of data for things like insurance actuary tables or
for the census every ten years. Information was so new, in
fact, that Professor Alex Rathe claimed, "as late as 1946 there
were in the combined professional, technical and scientific
press of the United States only seven articles on the subject
of information."[1]

Information today, of course, is taken more seriously; it
has become an essential commodity. "Theories based on the
concept of 'information' have so permeated modern culture
that it is now widely taken to characterize our times."[2] Yoking
such a paucity of research on information with the word

Bob Everett at Aircraft Stability and Control Analyzer, circa 1945.
(Photo courtesy of MITRE Corporation.)

"technology" seemed beyond even the keenest intellects of the day. A retrospective on business management for *Duns Review* in 1958 would point out: "Only in the past dozen years [since 1946] has the concept of information—as distinct from the papers, forms, and reports that convey it—really penetrated management's consciousness. That it has done so is largely due to recent breakthroughs in cybernetics, information theory, operations research, and the electronic computer."[3] Both Norbert Wiener's seminal work on cybernetics and Claude Shannon's breakthroughs on information theory came out in 1948, yet they did not wash through the streets of Cambridge until 1949; operations research burst out with the Chain Home system in the early 1940s; and digital computing's first electronic machine had just cranked into existence in 1946 at the University of Pennsylvania's Moore School of Electrical Engineering.

It's astounding to think that these twentysomething bright boys would by 1948 have invented the first real-time, electronic digital computer and by 1950 have wrapped their machine in cybernetics, information theory, and operations research, what we perceive today as the basic trappings of Information Technology. But that's exactly what they did. And in the process of all that wrapping and putting forth, they would also pioneer the new discipline that would make Information Technology universal—systems engineering. In 1946 the only thing in the bright boys' sights was their machine. Any realization about information—its ubiquitous power and future influence—was dormant, but not for very long.

In 1946, barely two dozen people in all of the United States knew what an electronic computer was all about, and none of the bright boys were among those elite few. In February of that year, Forrester had convinced the United States Navy that he could build one, and by March he had a contract in hand to do just that. Forrester's proposal called for the building of an electronic digital computer and a flight simulator/analyzer for $2.4 million to be delivered by 1950. The Navy accepted the proposal, funding the construction for an initial $1.2 million through mid-1948.[4] It was quite the sales job by the young engineer; Raytheon, a company that Vannevar Bush helped to found, offered to build a computer for the Navy for $650,000. Forrester persuaded the Navy to fund his electronic offering, saying that "the potential of the digital computer was so great and the benefits derived from its use so immense that the costs involved, no matter how great, were warranted."[5]

The Navy bought into the plan, naming it Project Whirlwind. The name Whirlwind was selected following the Navy tradition of naming its computers after air movements: Hurricane, Typhoon, and Zephyr were other

Navy computers. The bright boys were set loose to construct their electronic digital computer. First item on their agenda was to learn a thing or two about electronic computers—and to learn really fast! To catch up, they packed their bags for a trip to the Moore School for a series of lectures from the best electronics experts of the day. They hoped that something would rub off, and soon. As Ted Williams casually and confidently loped for home plate, the bright boys were about to grope a few tentative steps toward first.

Today, each time we step off an elevator or a subway train, or book an airline ticket, or pull some cash from an ATM, or even thaw a steak in a microwave, there is an unseen thread of discovery at work that leads directly back to the bright boys. In their old, brick laundry building on Massachusetts Avenue, the bright boys opened wide the digital doors to the future. Today, any modern gizmo that operates with ones and zeros has a lineage that harkens back to their inner sanctum of discovery and innovation. On their way to Pennsylvania, their wildest dreams could hardly have imagined such an outcome. And most assuredly, none ever reckoned that a bleacher fan in seat #21 might one day beam up the game's box scores to a friend in a passing jumbo jet—and that their work in the Barta Building might, in some small way, have something to do with it. In the summer of 1946, nothing was for sure for the bright boys. It would have been difficult just to foresee even near-term surprises like George Valley, Gordon Saville, the air defense of North America, and the apotheosis of real-time command and control waiting in the offing. But they all most surely were.

And that it would all one day come to form the bedrock of Information Technology—a dozen years (1958) before the term "Information Technology" was coined—was for Tom Corbett's eyes only. The bright boys would design and build the first real-time, electronic digital computer, the first digital network—replete with digital packet technology and modems over long-distance telephone lines—and soon thereafter, a veritable world's first Information Technology company, all stuck behind the walls of the Barta Building.

Even their business plan turned out to be as innovative and slickly modern as their advanced technology. They would design, build, and test all prototypes themselves and then outsource everything, all the while keeping a tenacious eye out for exacting quality over each and every aspect. The bright boys would lead a host of high-tech "wannabes" into the digital world; names like IBM, Bendix, Sylvannia, AT&T, Western Electric, and Burroughs got a shot at outsourcing for Whirlwind. And along the way, Forrester and Everett would inaugurate the concept of the "team approach," making it virtually synonymous with the only way for modern, high-tech wizardry to be truly successful.

Ironically, it all began with a job that someone else did not want to take on. In 1944, Bell Telephone Laboratories turned down a job, and in so doing, unwittingly sparked a revolution that jump-started the bright boys into history. Captain Luis de Florez, head of the Special Devices Division of the Office of Research and Inventions of the US Navy and an MIT graduate, took the job that Bell Labs had rejected and walked it into Gordon Brown's Servomechanisms Lab at MIT. Brown readily agreed to take it on. He placed the job into the gifted hands of the then 26-year-old Forrester and his equally gifted 23-year-old sidekick Everett, who together proceeded to set into motion a magical adventure in engineering history.

In 1944, Forrester and Everett assembled a group of fellow engineers and began to build a flight simulator for the training of Navy pilots. The Navy approved a preliminary design study for $75,000 for what was to be called the Aircraft Stability and Control Analyzer or ASCA. Prior to the ASCA, flight simulators were built for each and every type of aircraft; this time, however, the Navy sought a single flight simulator that could be used as a general-purpose trainer for any aircraft. In the summer of 1945, MIT received the Navy's contract for $875,000, although some Navy engineers fated it to early failure, calling the project "an engineer's nightmare."[6] As Forrester recalls, "It was to be rather like an aircraft pilot trainer, except that it was to be so precise that instead of acting like a known airplane, it could take wind tunnel data of a model plane and predict the behavior of the airplane before it was built."[7] The cockpit of the ASCA presented a difficult engineering problem, but it was the controller that was the real challenge.

The controller, the device that did all the necessary calculating and analysis, would need to be a high-speed, real-time computer. In 1945, high-speed computation meant only one thing: an analog computer, which was a colossal but intricate maze of metal gears, cams, spindles, and shafts. Quite alien looking to the modern eye, the room-sized machines looked like the giant handiwork of a clockmaker gone mad. But to contemporary engineers, they were marvelous machines of precision and eye-fetching wonder. Some were electromechanical, but even electricity could not power the gears beyond their limitations. Some, like Howard Aiken's Mark I at Harvard, were both electromechanical and digital but were still awkward performers. Today all computers are electronic—gearless and quiet with an internal cityscape of solid-state parts covered over with a metal box. The term *digital computer* is now synonymous with electronic and real time—but not so in the mid-1940s. Forrester and company began to build their computer as an analog machine. By the fall of 1945, all were ready to throw in the towel.

The ASCA system was frustratingly complex as a cumbersome analog gear works, and Forrester was unable to cope with implementing the simulator's aeronautical equations in real time. Fortunately, along came Perry Crawford. Crawford was a good friend of Forrester; an MIT graduate student who had done his 1942 master's thesis on automatic control by arithmetic means. Perry Crawford, Forrester remembers, "was a person with continually unfolding visions of futures that others had not yet glimpsed. He was always looking, listening, and projecting new ideas into the future."[8] Far from being the stereotypical researcher holed up in his lab, Crawford "was uninhibited, not restrained by protocol or chain of command, and a freewheeling intervener in many circles of activity."[9] And as luck would have it, the "uninhibited" Crawford, the digital visionary who pushed the idea of combat information and control by electronic computers, was, by October 1945, working for the Navy's Special Systems Division (which had taken over for the then recently defunct Special Devices Division) and now had oversight of the Whirlwind project.

Crawford "intervened" with Forrester regarding the potential of electronic control for his reluctant device. To his credit, Forrester quickly agreed. The unyielding technology had to be ditched, a new plan for success devised for the customer, and momentum maintained for MIT management as well as the project's 100-plus technicians who looked to Forrester for leadership. If Forrester ever pined for a smattering of digital reality, he need look no further than his attendance at Raymond Archibald's *Conference on Advanced Computation Techniques* held at MIT the last three days of October in 1945 where, among other things discussed, was that of the great ENIAC, purring away at the Moore School of Engineering at the University of Pennsylvania. Forrester had heard enough; he had to see what an electronic digital computer looked like up close and personal so he made a quick trip to the Moore School to kick the tires on the giant machine.

And giant the massive machine most surely was. It was eight feet tall and 80 feet long. It had 70,000 resistors, 10,000 capacitors, and a half-million soldered joints. Its 30 tons took up 3,000 cubic feet of space, and when its 17,468 vacuum tubes powered up, the room temperature jacked up to 120 degrees, and rumor had it that lights in sections of Philadelphia would flicker. The result of a $400,000 contract with the Army in 1943, the computer's purpose was to calculate gun-firing tables for the Army's nearby Aberdeen Proving Grounds. The trajectory of an artillery shell like the 155-millimeter "Long Tom" covered over 500 sets of conditions. A team of women punching numbers into mechanical calculators would take more than a month to calculate a single table—a table without which the gun could not be properly aimed and fired. But the hulking computer was indeed fast, over 1,000 times faster than any mechanical device. A trajectory that formerly took 20 hours to tabulate on a desk calculator, ENIAC could kick out in 30 seconds.[10] *This*, Forrester must have thought, changes everything. He was staring the future—his own future as well—dead in the eye. As *The New York Times* reported at ENIAC's official unveiling on February 14, 1946, the world's first electronic digital computer, ENIAC—or Electronic Numerical Integrator and Computer—is "an amazing machine which applies electronic speeds for the first time to mathematical tasks hitherto too difficult and cumbersome for solution."[11]

Exactly what the ASCA project needed. While hobnobbing at the Moore School, Forrester discovered upcoming plans for an even newer electronic machine called EDVAC (Electronic Discrete Variable Automatic Computer). As a consultant on the ENIAC project, the renowned mathematician John von Neumann had come up with a document called "First Draft of a Report on EDVAC." EDVAC's machine design was superior to ENIAC's, because EDVAC introduced a revolutionary idea, that of the stored program concept. "Stored program" means that the computer's memory possesses not only data, but also the instructions that the computer uses to manipulate that data. ENIAC's instructions were programmed by plugging wires into the machine and then having to rearrange the wires with each change in programming. A simple program change could take hours.

John von Neumann revolutionized that in a hurry. His report would become a seminal document in the history of digital computing. Every computer today is a stored program computer, and the innards of every computer are known as von Neumann architectures. The stored program concept "is still considered the most profound invention in, if not the very foundation of, modern computer science."[12] Forrester made sure that he got his hands on a copy of the report. He and Everett had in mind to build a machine bigger, better, and faster than either ENIAC or EDVAC. The first step in the conversion process would be the risky maneuver of switching the Navy from analog to electronic digital. The Navy might well have balked, finally agreeing with its own engineers who had early on cursed the project as impossible. Presenting the Navy with a good alternative plan, a little brinksmanship with some able help from Crawford, and together they prevailed.

It seemed that the Moore School was to be the epicenter of electronic computing and quite possibly held the key to Whirlwind's future success.

To be there for a special course offering on electronic computing was an absolute necessity. With the Navy's blessing for Whirlwind in hand as well as a fatter contract, and with MIT's concurrence, Gordon Brown's youthful digerati headed south in the summer of 1946 to attend the course on the new field of electronic computer design.

It would turn out to be an eventful journey filled with discovery and enlightenment. Held from July 8th to August 31st, the course, *The Theory and Techniques for the Design of Electronic Digital Computers*, was both attended by and taught by a veritable Who's Who of the founding fathers of modern digital computing; all of whom, except for John von Neumann, were great unknowns at the time. For Forrester and Everett, two young guys struggling to get their electronic creation off the ground, this was an important intellectual bazaar filled with new ideas, insights, and techniques all handed up by the originators of an onrushing digital future.

The bright boys were in the right place at the right time; and they had the right project with the right budget. Their digital learning curve would be traversed at the Moore School like a two-month apprenticeship in a medieval guild or the Bauhaus of the 1920s. As Forrester reminisced years later, "With the absence of the technology there was also the absence of thousands of people out there feeling that they knew how to do it better."[13] The bright boys were free to learn, to ponder, and to invent on their own. For Forrester and Everett, it was the beginning of a ten-year working partnership, and a great friendship that would span decades.

If two partners in business together are both alike, then one of them is totally unnecessary; so claimed Harold Geneen, former Chairman of ITT, founder of MCI, and a pretty fair judge of people in partnership. Although very different men from very different backgrounds, Forrester and Everett shared in exact proportions the same discipline, single-mindedness, and commitment to their beloved machine. Norman Taylor, another of the Barta Building bright boys, would say of them:

> Bob Everett was relaxed, friendly, understanding and I have never seen anyone who could go to the heart of a problem so fast! Jay was as fast, maybe faster, but he was always more formal, more remote somehow, and you weren't always sure how dumb he thought you were, or how smart. That kept us on our toes, I suppose. It was difficult to know what he was going to do next, but he was so terribly capable, it didn't matter if you couldn't follow his reasoning. He was always thinking with seven-league boots on. It made him a pretty formidable guy to work for partly because he and Bob always made sure you understood the problem you were working on, by finding out what you didn't know

as well as what you did know . . . I never resented Jay's obvious ability, but he wasn't the sort I'd call easy to work for. He definitely was never 'one of the boys.' He was the chief, cool, distant, and personally remote in a way that kept him in control without ever diminishing our loyalty and pride in the project.[14]

Jay Wright Forrester was born in 1918 into the great mid-West, specifically that northwest chunk of Nebraska known as the Sand Hills, a huge, swirled expanse of ancient sand dunes topped over with a thin crust of topsoil and grass. Grass, grass, and more grass for as far as the eye could see across Custer County; hundreds of square miles of it filling everything between the North Platte and Niobrara Rivers 20 miles from the nearest town of Anselmo. Just right for cattle ranching. An ocean of rolling grasslands also just right for the sojourns of a young boy's imagination. Unlike his rancher father, Duke, Jay had a decided mechanical bent and put it to good use on all manner of helpful contraptions, like an electrified screen door to zap bugs or, in his senior year of high school, a wind-driven system—uniting the open expanses—that brought the first-ever electricity to the farm house: light bulbs, radio, even a toaster were then possible. It may not have been the Tennessee Valley Authority, but to a youthful sense of accomplishment, it was mighty close. College took him clear across the state to the University of Nebraska at Lincoln. Tuition was $35 a term, and 1939 saw him graduate with a degree in electrical engineering. Quietly reserved, introspective, and handsome, the young Forrester was all of *Richard Cory* "clean favored and imperially slim" and soon found himself on the streets of Cambridge as an MIT graduate student. Goodbye Nebraska.

Nearly 1,500 miles from the Nebraska Sand Hills but just ten miles from the neatly trimmed outfield grass in Yankee Stadium, is the decidedly Eastern melting pot of humanity on the Hudson River—Yonkers, New York, the birthplace of Robert Rivers Everett in 1921. For another lad with a mechanical bent, Yonkers had the perfect atmosphere for inspiration into the engineering side of things. The world-famous "electrical wizard," Charles Proteus Steinmetz, electrified the Yonkers' trolleys; Edwin H. Armstrong invented FM radio broadcasting there; Leo Baekeland chose Yonkers to cook up his first batch of synthetic plastic, which he named after himself, Bakelite; while the first practical dynamo and first practical electric engine were both invented and produced in Yonkers. Easygoing, affable, and very bright, the sociable young Everett journeyed to Duke University, graduating first in his class in electrical engineering, immediately following which he set himself down at MIT for more of the same.

Both Forrester and Everett were fortunate to find themselves in 1942 as members of Gordon Brown's Servomechanisms Laboratory, which was a next-door neighbor to MIT's microwave wonder factory, the Radiation Laboratory, known simply as the Rad Lab. Forrester helped Brown to found the Servomechanisms Lab in 1940, becoming co-director the same year. (The neighboring Rad Lab began operations in June, 1940). Everett was a graduate student. In 1939, Brown, an Australian transplant to Harold Hazen's Department of Electrical Engineering, taught the school's first-ever courses on servomechanisms to Navy fire-control officers. The term servomechanism originated in the 1860s from the steering engine developed by Jean Joseph Farcot, which he called "servo-moteur." Servo devices were used to move naval guns into position for firing. Receiving an electrical instruction to move to a certain position, servos readily comply, and if they go too far, they automatically correct themselves. Control by the numbers was impossible until the advent of servo systems. With a world war raging, servomechanisms became vitally important to both the Navy and Army as well as for antennas for Rad Lab radar systems. It was a perfect opportunity to initiate a laboratory devoted entirely to servo devices.

Servomechanisms that were combined with tracking radar and anti-aircraft guns were especially important. Without a servomechanism it took 14 people to detect an aircraft, calculate its future position, and then to precisely move a gun, aim, fire, and hopefully hit the moving aircraft. Against fast-moving fighter planes, such a manual system was woefully inadequate. The famous SCR-584, developed by a Rad Lab team led by Ivan Getting and Louis Ridenour, successfully combined tracking radar, a mechanical computer, and servomechanisms for a deadly accurate anti-aircraft system.[15] Used against 104 V-1 rockets fired by the Germans at London, the SCR-584 shot down 100.[16]

The close proximity of the Rad Lab filled with physicists working to develop microwave radar also helped to broaden Brown and his young engineers' understanding of pulse circuitry.[17] Getting, a physicist who claimed to be the second or third employee hired at the Rad Lab, said that 90% of the 2,000 professional staff there were also physicists. The nature of working with microwaves was the reason for their great numbers because "a physicist's education in the fundamentals of electromagnetic theory and associated subjects . . . was more widely embedded than it was in the professional electrical engineer of the time. Professional electrical engineers generally spent time working on sixty-cycle power equipment or power transmission or transformers, or worked in radio waves modulated at sound frequencies, say up to ten kilocycles. They were not versed in microwaves at all."[18]

Such a handy dose of experience would go to benefit Forrester and Everett's later design efforts with Whirlwind. Everett, who worked on radar antenna mounts at the Servomechanisms Lab, later recalled that, "radar technology was coming out of World War II activities in tubes and pulse circuits and storage devices that had been developed for moving target indicators. The technological foundation of pulse circuits made possible electronic digital computers."[19]

Projects like the SCR-584 also brought into sharp focus the need for "teams" to complete complex machines: interdisciplinary teams of physicists, mechanical engineers, and electronic and electrical engineers were necessary to pool their talents and expertise. Getting's SCR-584 automatic tracking radar required servomechanisms that did not yet exist. It required engineering of very skilled, very specialized mechanical antenna mounts, which Forrester and Everett built while working with the Rad Lab's physicists. And they learned as well the ins and outs of high-frequency pulse circuits "more in the television type of one to ten megacycles," said Getting. "You would find it difficult to run into more 'run of the mill' electrical engineers, who had designed circuits and abused vacuum tubes way beyond their specifications."[20]

This on-the-job teamwork would bode well for Forrester and Everett when much later they needed to coax optimum performances from Whirlwind's thousands of chancy vacuum tubes. Team members also directly shared their knowledge at group sessions designed to spread evernewer technologies as soon as they sprouted in either or both labs. Guy Stever recalls one talk in particular where he broached a new use for radar that he had just carried home from England. "My theme was that it was time for radar to go on the offensive rather than being used in the main to defend against attacks . . . and that the British were using [it] for offensive sweeps over the Continent."[21]

This use of radar as an offensive weapon would become, once it was partnered with a real-time computer, the key element in the development of real-time command and control. And for theorists like Gordon Saville, it was the ideal way to push an electronic frontline close in on the enemy. Specifically, Stever noted that the microwave tracking radar developed at the Rad Lab would be much superior to the long-wave radar being used by the British. Stever's update on British radar sparked a chord with the Rad Lab crew, so much so that they asked Lee Dubridge (the Rad Lab director) to present a similar update on the state of American radar, which he "immediately asked Louis Ridenour to do at the next colloquium."[22]

This team approach and cross-fertilization of ideas were not lost on the young Forrester and Everett. Although the Rad Lab closed for business in 1945 and its staff quickly went their individual ways, its early-on dabbling in a systems approach to complex projects would be fully embraced and refined by the bright boys. Forrester and Everett transferred this wartime team concept directly into the Whirlwind development process. Recalled Forrester later: "We had a core team of engineers who had been through several cycles from basic research and on into development, design, production, and solving equipment failures in the military field. They knew each other's strengths and weaknesses and they understood the entire process of bringing an idea to field operation."[23]

The results bordered on the fantastic: their electronic digital computer would eventually average an unprecedented 99.8% uptime—and do so for a continuous 28 years.[24] Today, with computer models changing as rapidly as every six months and with computer crashes a frequent nightmare, the quality and longevity of Whirlwind-developed technology is still without equal. And the systems approach that they methodically refined soon became the crown jewels of engineering, and in some manner or form was purloined by every military or commercial organization with a complex project on its hands. "I doubt it would be possible today," commented Forrester on his team's success, "to design a system of that complexity in so short a time and at so little cost."[25] Yet, exactly time and cost would soon be the cudgels used by both MIT and the Navy to beat Whirlwind nearly to the edge of extinction. However, in the summer of 1946, it was two young guys in the full headed not for troubling times but rather on the road to the Moore School and their destiny.

In 1946 Ted Williams was not the only one running for home, so too were 11 million soldiers, sailors, and airmen. Sadly, nearly 400,000 young Americans would never return to enjoy the summer of 1946 or any other season. President Truman was hard about the nontrivial task of getting the nation's war mentality, war economy, and warring populace back to normal. It was an agonizing process for the military, which would go to extremes and then rapidly have to correct itself—just in the nick of time— or, for those marching off to Korea, just a tad too late. Defense spending in 1945 was $81.5 billion, then $44.7 billion in 1946, tumbling to a low of $13.1 billion for 1947.[26] Wartime rationing was over and 15 million radios were produced, keeping all the new arrivals informed about Vaughn Monroe's latest on *The Hit Parade* or about a whole lot of babies that were being born in what was to be the first year of the Baby Boom. There was even a manual produced on how to raise all these kids: Dr. Benjamin Spock and Pocket Books put out the 25-cent bible on child rearing, *The Common Sense Book of Baby & Child Care*, which eventually sold 50 million copies in 39 languages.

Radio was about to lock horns with the new media upstart, television. Families crowded around miniscule five-inch black and white screens for Ted Mack's *Amateur Hour*; or the first-ever TV soap opera, *Faraway Hill*, on the DuMont Network; or sports like the Joe Louis–Billy Conn heavyweight bout beamed out to 100,000 on *The Gillette Cavalcade of Sports*. Jobs killing the enemy for just over $70 a month had been plentiful, but for returning GIs jobs were mighty scarce at home. Those who had work averaged $2,500 annually (about $208 a month—some $138 more a month than for their battlefield heroics), and for half that salary anyone could pick up a snappy Plymouth sedan and fill the tank for 15 cents a gallon. In New York City, the Beat Generation kicked off when Jack Kerouac and Allen Ginsberg first encountered the charismatic Neal Cassady, who inspired both of their in-fluential literary classics, *On the Road* and *Howl*.

Although the first 30,000 home air conditioners rolled out of America's factories to beat that summer's heat wave, an air-conditioned movie theatre was the best bet for a few hours respite. Millions of Americans did just that. In record numbers not seen since *Gone With the Wind*, moviegoers jammed in to catch the year's cinema blockbuster *The Best Years of Our Lives*. The Samuel Goldwyn classic about three returning servicemen in the aftermath of World War II, based on a *Time* magazine pictorial from 1944, explored the troubling irony that for many servicemen "the best years of our lives" were spent overseas in wartime and not upon their return home.

The starkly realistic and poignant film won the best of nearly everything at the Academy Awards, grabbing seven Oscars. Stiff competition at the 1946 Oscars came from other soon-to-be classic films, *The Razor's Edge*, *The Yearling*, and *It's a Wonderful Life*, which all seemed to explore the un-derstandable post-war themes of soul searching, introspection, self-sacri-fice, and right doing. Ferried to their hometown in the bombardier's space of a B-17, the three veterans in *The Best Years of Our Lives* look down on a graveyard for bombers. Fred, a former B-17 bombardier (played by Dana Andrews), is stunned: " . . . they're junking them . . . Boy, oh boy, what we could have done with those in '43! Some of 'em look brand new, factory to scrap heap. That's all they're good for now." Soon to get chucked upon the very same post-war scrap heap would be analog computing, and the Moore School lectures would have a major hand in lifting it to the very top of the pile. It was out with the old and in with the new, and the bright boys' plan for Whirlwind would help to usher in a new age of technology. They were on the road to Pennsylvania, and, just like the rest of post-war America, things would never be quite the same again.

The Day the Gears Stopped

The death knell for analog computing was the sound of its own gears grinding to a halt. Analog gears were on the electronic chopping block as were most other moving parts in analog computers and calculators. The Moore School's 48 remarkable lectures on going electronic were telling for their absence of anything relating to gears. And everyone attending the Moore School lectures knew it and had come to witness, among other things, the first jarring screech in the slowdown, even though all had been reared in a world of gears and made their livelihoods from it. Hastily organized two months after ENIAC's coming out party and financed by Army and Navy Ordnance, the Moore School lectures were by invitation only to 28 individuals. The military had witnessed first hand the promise of electronic digital computing and now wanted to push forward the development edge with the hope of getting more of the same or better. Maybe by financing lectures to a hand-picked audience a few new gems might soon glitter into existence.

It is doubtful that without ENIAC the Moore School lectures would have happened at all. ENIAC had charisma and its success had world-class pull. Although monstrous and flawed, ENIAC was real. It was not a rack of electron tubes hidden away in a research lab or a half-baked prototype gone half wrong or even a theoretical paper with an eye to the future. ENIAC was as big as life; you could pet it. And it worked. In a very real sense it was like any other successful commercial product: it was bought and paid for by a customer, worked to the customer's delight, and now the customer wanted more.

Thomas Edison's movie camera was a beast of a thing but it was a first and it worked. No one remembers or cares about William Friese-Greene's second-place offering or the legion of wonderful technical innovations that paved the way to Edison's invention. The movie camera forever changed how people record and acquire information. In the long run that is all that history really remembers: events and things that reshape the world. ENIAC was just such a change agent. And everything that preceded it, no matter how inventive, innovative or a stretch of genius on the road to electronic computing, paled in its shadow. ENIAC was the Cro-Magnon to the analog gear work's Neanderthal: a true departure from the road previously mapped out as the way to computing. ENIAC pushed into being one of those signal events when technological change is so startlingly new that

Marshall McLuhan says the entire *sensorium* of the brain needs to turn and adjust itself to properly comprehend. *Sensoria* would take some time to reorient to the full electronic drift.

Breakthrough ideas about the relationship between computing and the human intellect, like computer and Internet pioneer J. C. R. Licklider's "man-computer symbiosis" would not appear until the early 1960s. This was the dawn of electronic computing; its Iron Age of huge, hulking cabinets crammed high with components, thousands of hot electron tubes, and miles of hand-wired connections. And ENIAC was its first lord of the iron. Without ENIAC and its success, the Moore School lectures would have been hard pressed to be anything else but more visionary ramblings on what the coming electronic future of computing might portend. ENIAC was electronic computing's Big Bang; an impact heralded as big stuff on no less a public stage than the *New York Times*.

ENIAC commanded attention and its prowess commanded attendance at the lectures. A team of Moore School staff led by two no-names, physicist John Mauchly and his youthful engineering pal J. Presper Eckert, had painstakingly built ENIAC into what computer pioneer and historian Herman Goldstine would later call the most complex machine ever built up to its time. It was now up to 28 Moore School invitees to take up the challenge and to go beyond. Forrester and Everett were but two—but a very important two. Whirlwind would in short order change everything that preceded it and hurry into being the integration of computing machines and people into everyday life.

From antiquity onward gears had been synonymous with engineering, even as popular fascinations for artists who routinely depicted engineers with a familiar gear or two nearby. A gear works presented a computer engineer with a comforting place of notched wheels, drums, spindles, and axels; things that could be seen, felt, and heard as the gears churned at their tasks. How could it be that analog gears were coming to an end? For sure, gears were important and necessary things. Or were they?

The grinding to a halt was inexorably complete over a remarkably short period of time. Such a brief lifespan gives pause for thought about their true identity. Were they really computers or, as Maurice Wilkes suggests, merely calculators—calculators grown gargantuan as they approached the limits of their capabilities? Whatever their true identity may be, in 1946 they were technically known as analog computers. These analog giants came into prominence in the 1930s, and by 1947 were readying themselves for museums. Vannevar Bush's world-famous Differential Analyzer from 1932, the pride of MIT's computer engineering prowess, was a gear-laden analog machine that solved differential equations by factoring simpler lin-ear equations. In giant size, its innards resembled those of many conjoined grandfather clocks with oversized gears, integrators, sliding racks, cams, springs, and drive shafts all married together to fill the floor of a warehouse. The Moore School had two of them, both in constant use during wartime to calculate artillery-firing tables for the Aberdeen Proving Grounds.

What ENIAC accomplished in 30 seconds a differential analyzer took 15 minutes to perform. And that was just for small jobs. ENIAC's first big post-war job was a calculation in nuclear physics that would have taken 100 human years to solve by conventional methods. ENIAC solved the problem in two weeks (two hours of which was for the calculation, the bal-ance spent in checking the results), and ENIAC was far more accurate. Any answer going out lots of decimal places was risky business for an analog computer, but for an electronic digital machine it was a piece of cake. And therein resided most of the reason for its demise: speed of calculation and increased accuracy every time out was the post-war order of the day. And faster still was more like it.

By today's standards, ENIAC's two hours or even thirty seconds were way too long to assist something like jumbo jets stacked up by the doz-ens over Chicago's O'Hare Airport and landing at the rate of one every two minutes, or for banks needing to accurately process billions of checks daily. Each demands speed and accuracy—and to do both in real time. Whirlwind would be all three at once: fast, accurate, and in real time. And that was the electronic siren call of the Moore School that sent the youthful Forrester and Everett winging south in the summer of 1946. The careers of analog engineers by the thousands, those that could not or would not ac-cept the new world order and retrain themselves for it, met untimely deaths along with their machines. "The rise and fall of this profession," sadly re-called George Valley, "is a poignant story—of expectations that came true for other people, of ruined careers, of competent engineers pushed down to technician level."[27]

By contrast, the Moore School's electronic digital world was unseen and untouchable—everything happened deep within a box. The discovery of the transistor was yet another two years' off, still further out were integrat-ed circuits and modern computer memory. Back then, memory for data, and later, commands, was held in strange places like mercury delay lines or magnetic drums or electron tubes. Getting commands into ENIAC—programming—was accomplished by arranging a set of wires to one end of the machine. Switching on the electricity—a lot of electricity—would then provoke magic to happen inside ENIAC's box of tubes and circuits, which then, in turn, quickly pushed an answer out the other end. All done without gears. Strange stuff for the comfort zone of an analog engineer to abide.

Although some electronic digital computers, ENIAC included, would be built to operate on the decimal system (our familiar base-10 number system), the real killer system that would win the day was the binary or base-2 system, which converts all decimals into binary digits of 1s and 0s. The decimal number 5, for example, binary notation would translate as: 00000101. The two binary digits 1 and 0 would prove to be ideal for designing digital logic circuits that use two primary voltage states of "Hi" and "Low;" or as "Yes" and "No" for devising basic rules of logic for those same circuits; or when transistors eventually arrived on the scene, the simplicity of having only to switch back and forth from "full on" to "full off" was a perfect fit.

The language of every digital machine today is this same binary lingo. Press the number 5 on a keyboard and a digital machine, using a simple set of on/off commands, converts the decimal for 5 into a series of binary 1s and 0s that read: 00000101 in the machine's native tongue. Press the plus sign on the keyboard to request the operation of addition, followed by the decimal 7, which translates into binary as 00000111, followed by pressing the "Enter" key and the machine complies with the computation of 00001100, which it instantly converts for human consumption as the decimal number 12. The alphabet is similarly attended to by binary notation: the letter A is 01000001, B is 01000010, and C is 01000011, and so on.

Answers produced by ENIAC had no easy visualization like a monitor or laser printer to display themselves. Rather, it used a clunking contraption called an IBM card punch that literally punched holes representing the numbers into a card from which an IBM Tabulator would then read the holes and print the corresponding numerals. Entering numbers into ENIAC was just as laborious and nonintuitive. The bright boys would be the first to use a keyboard to input to a computer and the first to output to a display screen.

Operating at a rate of 100,000 pulses every second, ENIAC's accumulators were capable of adding 5,000 numbers a second. Breathtaking then, a pittance now. By comparison, a modern desktop computer with a relatively slow 500-megahertz chip performs a half-billion calculations every second. And that's from a machine that fits on a desktop; ENIAC filled an entire room. Also unlike ENIAC, every computer today must be a real-time machine, because no military commander, no air-traffic controller, no bank, no insurance company, no merchant of any kind, not even junior doing his homework waits for answers.

In 1946, no one wanted or expected a computer to spit out answers in real time; they were thankful for a two-hour wait with ENIAC, which was far preferable to holding hands with a calculator for 100 years. ENIAC was plenty fast for them and cracking tough equations was enough of a result. Computers doing real jobs like tracking a few enemy bombers or automating a factory or trading a billion shares a day at the stock exchange just did not compute. The sensible kids from the Sand Hills and Yonkers, who had spent a war building complex yet practical machines for desperate times, knew better.

ENIAC pointed the way to the bare essentials of electronic computing success but still had major development issues. And *real-time* electronic computing was absolutely nowhere to be found on the physical or mental landscape of its pioneers, but its need would soon become absolute. An amusing portrait of electronic computing *without* the advantage of real time is starkly evident in Licklider's influential paper, "Man-Computer Symbiosis." Writing in 1960 about the aims of computing, he says, another aim

> . . . is to bring computing machines effectively into processes of thinking that must go on in "real time," time that moves too fast to permit using computers in conventional ways. Imagine trying, for example, to direct a battle with the aid of a computer on such a schedule as this. You formulate your problem today. Tomorrow you spend with the programmer. Next week the computer devotes 5 minutes to assembling your program and 47 seconds to calculating the answer to your problem. You get a sheet of paper 20 feet long, full of numbers that, instead of providing a final solution, only suggest a tactic that should be explored by simulation. Obviously, the battle would be over before the second step in its planning was begun.[28]

That Forrester and Everett even contemplated building a real-time electronic computer—especially as early as 1946—should have been cause for a great stir of support, especially from the military. And it was, but the initial hurrahs would be short lived. ENIAC, begun at the Moore School for the Army in May of 1943, was finished in October of 1945, and then publicly unveiled in February of 1946. Much of its design work pre-dated 1943 by two years. EDVAC proposed by von Neumann in 1945 would not be completed until 1953. Forrester, for his Whirlwind machine and simulator, had promised the Navy a finished product by 1950; an aggressive estimate in light of ENIAC's timetable of nearly five years. Given that known spread in time for the design and construction of a large-scale, electronic digital computer, it would have been safe and reasonable to assume that Whirlwind could be built somewhere in the close vicinity of four years. Adding "real time" to the electronic schedule might well take a bit longer.

Unfortunately, time for the bright boys would be very ungracious—even a speedy two years was too long for some. Time would be more like a rug always threatening to pull itself out from under them. Standing on it while simultaneously trying to build a revolutionary computer would call for great balance and nimble resolve.

Old Lessons for a New Machine

The Moore School of Electrical Engineering at the University of Pennsylvania, 200 South 33rd Street, Philadelphia, July 1946. Every school kid knows the Wright Brothers' Kitty Hawk or Edison's Menlo Park. Mauchly and Eckert's Moore School might well have been elevated into that same pantheon for their birthing of modern electronic computing had it not been for greed, petty jealousies, politics, patent problems, muddled thinking, and bad judgment. Being a first can have its downside.

Young Forrester and Everett would not only have a front row seat to learn from the famous Moore School lectures but also a place ringside to witness the thrashing of a magnificent event in the history of technology. The Moore School enlightened the bright boys on the fine points of building an electronic computer as well as how to grapple with the gritty clinches of reality that might attend any or all of their future success.

The coming years might well have been quite rosy for the University of Pennsylvania had it not been for the digital rift cracking open in its midst. The graduate science programs and the Moore School of Engineering would have been direct beneficiaries of ENIAC's electronic prowess; and its Wharton School of business management as well as its renowned medical school would have been perfectly positioned for leadership roles as electronic computing inched forward into the realms of business and medicine. Philadelphia-area firms Remington Rand, Philco, Burroughs, and a division of Sperry were prime candidates for Wharton to finesse into digital futures. In nearby New Jersey, at the Institute for Advanced Study, was a heavyweight braintrust with the likes of Albert Einstein, John von Neumann, Kurt Gödel, and Alonzo Church to help out if the mathematics got a bit too rough. And also nearby at the Aberdeen Proving Grounds was ENIAC's deep-pocketed and very enthusiastic benefactor, the Army Ordnance Department's Ballistics Research Laboratory, which had already authorized the development of ENIAC's successor, EDVAC, in 1944, which preceded the writing of the "First Draft of a Report on EDVAC" of June 1945. Things don't get much better than that, not even in storybook endings or Hollywood movies.

However, four days after the City of Brotherly Love celebrated its first post-war Fourth of July and the Moore School lectures were about to commence, the promise of good tidings for computing at the University of Pennsylvania were all

Moore School of Electrical Engineering, University of Pennsylvania, circa 1947. (Photo courtesy of Collections of the University of Pennsylvania Archives.)

but gone, and needlessly so. When a couple of no names like Mauchly and Eckert created a really big-name computer and the subsequent patent for that big-name invention could possibly mean big-time money, fame, and power for a whole lot of people other than the no names who invented the machine, the tranquil halls of academia can get downright goosey. Most researchers can strive for a lifetime and never come close to a major coup like ENIAC. Here was Mauchly and Eckert's, wrought by their labor and intellect, which just maybe was in jeopardy of being snatched away from them.

It certainly was not an uncommon fate to befall unknown researchers. An understandable foreboding was afoot at the Moore School; one made to feel especially real when the no names have movers and shakers like John von Neumann in the mix. Von Neumann, the Hungarian-born polymath and towering intellect, was being hailed as a modern legend. He was father of game theory, with further of his mathematical contributions going to set theory, ergodic theory, and operator theory, plus more of his originality lavished on quantum mechanics, logic, computing theory, and economics. If there were a Nobel Prize in mathematics, which there is not, he would have owned it. A big name on the Manhattan Project and frequenter to Los Alamos, confidant to President Truman, advisor to the Atomic Energy Commission, buddy to the Joint Chiefs, and a phone call away from the Secretary of Defense, von Neumann was a real heavy hitter.

An inkling of ill ease for ENIAC's creators may well have set in when von Neumann in December of 1944, together with a group of fellow geniuses, formed the Teleological Society and needed a computer to "join their grand conjunction of ideas."[29] Mauchly and Eckert were not on the membership list. When von Neumann wrote the "First Draft of a Report on EDVAC," to which Mauchly and Eckert had contributed but were absent from any attribution, it caused them deep resentment. ENIAC's project supervisor, John Brainerd, who understood little about the workings of the giant machine,[30] was accused of disingenuously submitting an ENIAC report to the Applied Mathematics Panel of the government's National Defense Research Committee, claiming more than just a supervisory interest in the computer.[31] Soon after the incident, Brainerd resigned as the leader of the new EDVAC project.[32]

Bickering and controversy followed the project like a dark cloud. Carl Covalt Chambers, who early on hired both Mauchly and Eckert into the Moore School, was a keen observer of the computer's development lifecycle and a first-hand witness to the ongoing imbroglio. Mauchly disliked Brainerd and vice versa; and the Brainerd-Eckert relationship was rancorous and distrustful.[33] "Eckert was," said Chambers, "so far ahead of the thinking of other people that he rankled people—particularly Brainerd."[34] Chambers, an organizer of the Moore School lectures and later dean of the school beginning in 1949, was a Moore School loyalist who tried to preserve what he could of the project. "I think the one who caused more trouble was . . . Goldstine, because Goldstine was trying to put his foot into things. The friction was strong between Brainerd and Goldstine, too; and to some extent between Goldstine and Eckert, because Goldstine tried to run things."[35] Surprisingly, ENIAC persevered. However, the machine's creators did not; Mauchly and Eckert responded by agitating for the patent rights to their invention and angling toward the start of their own commercial venture.

George McClelland, the president of the university, was much "distressed by the controversy that occurred in 1944 when Eckert and Mauchly filed for patent rights on the ENIAC and refused to assign the rights to the university."[36] Harold Pender, dean of the Moore School, persuaded McClelland to allow Mauchly and Eckert to proceed. In 1945 von Neumann's Los Alamos pals, Nicholas Metropolis and Stan Frankel, virtually commandeered ENIAC to set up and supervise the first set of nuclear physics calculations to be run on an electronic computer. Their computer runs were associated with top-secret studies of thermonuclear chain reactions for the US hydrogen bomb project. The first set of calculations took place in November of 1945, and subsequent calculations continued up through the time of ENIAC's formal dedication in February of 1946.

Goosey quickly turned to divisiveness. In 1946, with ENIAC splashed across the *New York Times*, Mauchly and Eckert were given an ultimatum by Irven Travis, supervisor of research for the Moore School, to sign a patent release or else. They resigned that March. A year later Travis left to become a vice president at Burroughs charged with starting up its computer division. That same June saw John von Neumann, Arthur Burks (another former ENIAC engineer), and Herman Goldstine (University of Chicago PhD in mathematics, former Army liaison officer to the Moore School as well as the guy who first introduced von Neumann to the ENIAC project), get together to draft a document called the "Preliminary Discussion of the Logical Design of an Electronic Computing Instrument" for the Institute for Advanced Study. Army money flowed their way for von Neumann's new computer, which was named the IAS computer after the Institute. Julian Bigelow, a Teleological Society member, was made chief engineer and worked on the project together with Burks and Goldstine.[37]

That August following the lectures, the Army dismantled ENIAC and carted it off to Aberdeen and its Ballistics Research Laboratory where it performed faithfully until final shut down in October of 1955. An independent computer company, Mauchly and Eckert's Electronic Computer Corporation, popped onto the Philadelphia business scene. Two years later, after six years of development, the EDVAC struggled into existence at the Moore School— without an input/output device. By then its experienced electronics engineers were long gone. Seemed like everyone got something out of the ENIAC affair except the University of Pennsylvania.

Benjamin Franklin, America's engineer extraordinaire and the one whose "Proposals Relating to the Education of Youth in Pensilvania" (1749) first suggested the founding of the school, must have quaked in his resting place at the botched attempt at a little glory for the old hometown. Since there was ample policy and precedent existing with government contracts, especially those with the Department of War, which would have satisfied everyone all around, it is amazing that ENIAC's halo lingered so briefly at the school. ENIAC was built under contract to the Army. As Edward Walterscheid points out in the *Harvard Journal of Law and Technology*, a title-granting approach called the "license" policy was standard procedure for such government work both during and after the war. The "license" policy would allow the University of Pennsylvania to control the patent with the ability to assign it, perhaps to Mauchly and Eckert, as long as a grant to the United States was made for a nonexclusive, irrevocable, royalty-free license to use the invention in other research endeavors, and the invention was placed into adequate commercial use by the university or its assignee within a designated period of time.[38] The government took this rather liberal "license" policy stance because it felt that it was "the most effective means of obtaining R&D work from the most competent contractors."[39]

McClelland, who was a humanities guy with little experience other than in academia, was wary about military support to the university in peacetime, which may have prompted him to vacate any interest in the ENIAC patent.[40] The "license" policy, together with a little good faith all around, might have made all the difference in simmering down any paranoia or ill will among the parties. Such a happy conclusion would have served the university well and also might have helped Mauchly and Eckert to dodge the eventual court decision in 1973 that voided their patent claim. However, after everything unraveled, the University of Pennsylvania's computing leadership never recovered. It would not be until 1950 that the Moore School even offered a course in electronic digital computing in its curriculum.[41] And Mauchly and Eckert, great computer makers but abysmal businessmen, would toil with incessant financial hardship. As Forrester and Everett strolled down Walnut Street to their first lecture, their first lesson in real-world computing had already taken place.

John von Neumann was far from being a malevolent character in a digital morality play. As Howard Rheingold describes him in *Tools for Thought*: "Everyone who knew him remembers to point out two things about von Neumann—how charming and personable he was, no matter what language he was speaking, and how much more intelligent than other human beings he always seemed to be, even in a crowd of near-geniuses."[42] He was highly principled and quite above filching anyone's work or invention, in particular that of Mauchly and Eckert, who presented him with a magnificent new means to do his calculations. Before his chance meeting with Herman Goldstine on the Aberdeen train platform in 1944, von Neumann worked with Howard Aiken's Mark I electromechanical machine. With ENIAC, he quickly realized that his mathematical work would have a computing tool that was 1,000 times faster than the Mark I.[43] Quite the upgrade for a mathematician to fall into accidentally.

This introduction to electronic computing provided von Neumann with the only way to find answers to previously intractable problems.[44] He was more than grateful for the opportunity and hell-bent on electronic computation assisting his work. The phenomenon of turbulence, the dynamics of explosions and implosions, and the modeling of complex weather patterns each "required such enormous numbers of calculations the future progress in the field was severely limited by the human inability to calculate the results of the most interesting equations in a reasonable length of time."[45] With an electronic computer plowing through equations for him, von Neumann could now arrive at a clearer understanding of the forces at work in these phenomena. What von Neumann really sought was what Licklider would also seek: to spend less time doing calculations and more time figuring out what they meant. When ENIAC could not fulfill all the answers to igniting deuterium for a fusion bomb,[46] von Neumann saw the need to construct his own machine, which he realized in the IAS computer.

The absolute necessity for electronic computers was firmly planting itself, while for analog machines it was just a matter of time. What did not change, however, was how the new machines were used; they continued to be equation crackers—only better and faster equation crackers. "For several generations," said Eckert in an interview years later, "people had known how to compute a lot of engineering stuff which nobody, in fact, was computing because it was too difficult to compute."[47] That's why von Neumann reached for ENIAC in a hurry. Electronic computers were to become quickly identified by the mathematics and physics problems that they helped to solve and, as such, were perceived as scientific computers only. A famous example of that narrow mindset comes from an inventive graduate student of von Neumann, who created a programming language that much simplified the tedium of computer coding in binary. Von Neumann quickly admonished the student, saying, "It is a waste of a valuable scientific computing instrument to use it to do clerical work." Many others, especially mathematicians, held the same myopic view of electronic computing as von Neumann. Or maybe they held the view merely because the Great One held it, because how could John von Neumann ever be wrong or even off base about anything relating to calculations? Such a strident and single-minded perception was reminiscent of the B-17 bomber clan and its affection for unescorted daylight bombing. It seemed that any nonscientific use of a computer was somehow improper. This specter of computers *only* as scientific instruments would come to haunt Whirlwind's progress and threaten its success. Real-time computing, real-time command and control—the rootstock of Information Technology—would depend on Whirlwind to beat the odds.

Both Cathedral and Bazaar

Around campus the Moore School building at 33rd Street and Walnut was known as the Whistle Factory. Some say the name came about because of the noises emanating from ENIAC, others because it was the former home of the Pepper Musical Instrument Factory from 1911 to 1926. From July 8, 1946 to August 31, 1946, it was the site of the first-ever conference on electronic digital computing. Officially it was an eight-week course: *The Theory and Techniques for the Design of Electronic Digital Computers.* At a time when there were very few conferences in science and technology in general and fewer still in computing and none in electronic digital computing,[48] this was a remarkable occasion for the existing digerati, of which there were few worldwide, to come together to pool their talent and expertise.

Most everyone was from the East and industry or university affiliated, some trickled in from California, and still others arrived from England. Getting anywhere in 1946 was arduous; one could not just jump a quick jet for Philadelphia. The closest thing to air travel was leapfrogging in a twin-prop DC-3. The US interstate highway system was still more than a decade off, but gas was cheap and the scenery not too bad on a journey south. That left buses and trains for long distances, or for hopping oceans, the lumbering DC-4. Still, 28 attendees came for the 48 lectures.

Each lecture was given in the morning at the rate of five per week, each lasting about three hours, which was followed every afternoon with an informal discussion on that morning's topic. "It was an opportunity to tell the world what we'd been doing," recalled Chambers, "and it gave a lot of people training."[49] The lectures were probably the only way for the Moore School to get the word out; published papers would be too slow in a rapidly changing electronic scene (the ENIAC's technology was already fading fast in favor of EDVAC's). Additionally, electronic computing had no learned journals in which to publish. Herbert Grosch, who was the leader of IBM's Watson Laboratory in 1945, said that "what professors call 'the literature' there wasn't any—at least for computers . . . the exception, *MTAC* (*Mathematical Tables and Other Aids to Computation*), published quarterly at Brown University. In the Thirties and Forties computer articles were turned away by engineering and scientific journals unless well sponsored, even *MTAC* was careful."[50]

The ENIAC team, divided and dispersed, came together in full force for the lectures. J. Presper Eckert delivered eleven of the lectures; John Mauchly six; Herman Goldstine six; Bradford Sheppard six; Arthur Burks three; Kite Sharpless two; and Hans Rademacher and Jeffrey Chu one each. John von Neumann contributed a lecture, as did Irven Travis. An impressive lineup of nine other future digital notables also contributed as lecturers. Forrester's good buddy, the irrepressible Perry Crawford was one, so too was the king of electromechanical calculation, Howard Aiken. The complete list of lectures and their topics displays the comprehensive approach taken by the organizers at the Moore School, "to tell the world what we'd been doing" and to "give a lot of people training" in the dark arts of electronic digital computing:

George Stibitz Bell Labs	Introduction to the Course on Electronic Digital Computers
Irven Travis Moore School	The History of Computing Devices
John Mauchly Electronic Control Co.	Digital and Analog Computing Machines
Derrick Lehmer Univ. Calif., Berkeley	Computing Machines for Pure Mathematics
Douglas Hartree Univ. Manchester (UK)	Some General Considerations in the Solutions of Problems in Applied Mathematics
Herman Goldstine Inst. of Advanced Study	Numerical Mathematical Methods I
Herman Goldstine Inst. of Advanced Study	Numerical Mathematical Methods II
Arthur Burks Inst. of Advanced Study	Digital Machine Functions
John Mauchly Electronic Control Co.	The Use of Function Tables with Computing Machines
J. Presper Eckert Electronic Control Co.	A Preview of a Digital Computing Machine
C. Bradford Sheppard Moore School	Elements of a Complete Computing System
Herman Goldstine Inst. of Advanced Study	Numerical Mathematical Methods III
Howard Aiken Harvard University	The Automatic Sequence Controlled Calculator
Howard Aiken Harvard University	Electro-Mechanical Tables of the Elementary Functions
J. Presper Eckert Electronic Control Co.	Types of Circuit: General
Kite Sharpless Moore School	Switching and Coupling Circuits
Arthur Burks Inst. of Advanced Study	Numerical Mathematical Methods IV

Lecture list courtesy of John R. Harris, Virtual Travelog.

Herman Goldstine Inst. of Advanced Study	Numerical Mathematical Methods V
Hans Rademacher Moore School	On the Accumulation of Errors in Numerical Integration on the ENIAC
J. Presper Eckert Electronic Control Co.	Reliability of Parts
C. Bradford Sheppard Moore School	Memory Devices
John Mauchly Electronic Control Co.	Sorting and Collating
J. PresperEckert and **C. Bradford Sheppard** Electronic Control Co. and Moore School	Adders
J. Presper Eckert Electronic Control Co.	Multipliers
John Mauchly Electronic Control Co.	Conversions between Binary and Decimal Number Systems
Herman Goldstine Inst. of Advanced Study	Numerical Mathematical Methods VI
Jeffrey Chu Moore School	Magnetic Recording
J. Presper Eckert Electronic Control Co.	Tapetypers and Printing Mechanisms
John Curtiss National Bureau of Standards	A Review of Government Requirements and Activities in the Field of Automatic Digital Computing Machinery
Herman Goldstine Inst. of Advanced Study	Numerical Mathematical Methods VII
Arthur Burks Inst. of Advanced Study	Numerical Mathematical Methods VIII
Perry Crawford Office of Research and Inventions, U. S. Navy	Application of Digital Computation Involving Continuous Input and Output Variables

J. Presper Eckert Electronic Control Co.	Continuous Variable Input and Output Devices
Sam B. Williams Bell Labs	Reliability and Checking in Digital Computing Systems
J. Presper Eckert Electronic Control Co.	Reliability and Checking
C. Bradford Sheppard Moore School	Code and Control I
J. Presper Eckert Electronic Control Co.	Code and Control II: Machine Design and Instruction Codes
C. Bradford Sheppard Moore School	Code and Control III
Calvin Mooers Naval Ordnance Laboratory	Code and Control IV: Examples of a Three-Address Code and the Use of "Stop Order Tags"
John von Neumann Inst. of Advanced Study	New Problems and Approaches
J. Presper Eckert Electronic Control Co.	Electrical Delay Lines
J. Presper Eckert Electronic Control Co.	A Parallel-Type EDVAC
Jan Rajchman RCA	The Selectron
Calvin Mooers Naval Ordnance Laboratory	Discussion of Ideas for the Naval Ordnance Laboratory Computing Machine
J. Presper Eckert Electronic Control Co.	A Parallel Channel Computing Machine
C. Bradford Sheppard Moore School	A Four-Channel Coded-Decimal Electrostatic Machine
Kite Sharpless Moore School	Description of Serial Acoustic Binary EDVAC
John Mauchly Electronic Control Co.	Accumulation of Errors in Numerical Methods

There was an attempt to memorialize all 48 lectures for later use. It was a grand thought, since so little publishing or conference support was accorded electronic digital computing in the early days of the technology. A published record of the lectures might well be useful to many.

Each three-hour, morning lecture was recorded using a wire recorder, which was a chunky, box-like machine atop which was inserted a three-inch diameter by one-inch spool of steel wire, much like fishing line. An ancestor of oxide-based magnetic tape, the thin wire was dragged over a recording head while the lecturer spoke. Each spool with 7,200 feet of wire at a recording speed of 24 ips resulted in one hour of audio recording. The plan was for each lecturer to go away with his wire recordings and notes and produce a manuscript for publication. It would have made for a nifty boxed set, yet didn't go as planned. Some of the lecturers followed through, others took what seemed like forever to produce a manuscript, and still others did nothing. Between late 1947 and June of 1948, four volumes were published. Some volumes did not fair too well, like Volume IV, which was and still is missing nearly half of its lectures.

The great takeaway for both the lecturers and attendees alike was what each brought to the others—a morale-boosting, intellectual camaraderie oozing with the firm conviction that they were involved in the creation of something truly magnificent. Within the drab confines of the Whistle Factory electronic digital computing was alive and well; a working ENIAC was there in all its glory, while plans and ideas were afoot everywhere for newer and better machines.

The Moore School lectures lifted electronic digital computing out of its isolation. While its collective intelligence poured out, its collective eye watched, evaluated, contributed, and grew. It was Eric Raymond's "cathedral" builders hooking up with "bazaar" builders to assess, refine, and fix their mutual electronic calling. Raymond's famous 1998 essay, "The Cathedral and the Bazaar,"[51] on the building and debugging of software, puts software development into two distinct camps. There are the cathedral builders, who are a tight-knit, dedicated few working through all of the software's problems and then releasing it to the world—the standard model for software development—analogous to ENIAC's core development crew. And then there is the bazaar-building model where software leaps out into the world and actively seeks after thousands of codevelopers—working online—to scrutinize, rework, and to suggest corrections. This was Linus Torvald's method for LINUX development—similar to the dozens of eager intellects gathered in the Whistle Factory probing each lecture and lecturer. Raymond's view is that "given enough eyeballs, every problem will be transparent to somebody."[52] In microcosm, the Moore School was both a cathedral and a bazaar for early hardware and programming that resulted in a much-needed outflow of new advances for an emerging technology.

And there were eyeballs aplenty in the Whistle Factory. Maurice Wilkes, over to attend from Cambridge University, cast a glance at the EDVAC's stored-program concept, then returned to England and built the first operational stored-program electronic computer, his EDSAC or Electronic Delay Storage Automatic Computer. Sam Alexander, chief of the Electronic Computer Laboratory for the National Bureau of Standards, was there with four of his crew, notably Edward Cannon, who would produce the US government agency's first electronic computer, the SEAC or Standards Eastern Automatic Computer. In August, an ENIAC part-timer, the American mathematics professor Harry Huskey, would leave for England and Manchester University to work with Max Newman and England's resident computer genius, Alan Turing, on another of electronic computing's firsts, the Automatic Computing Engine or ACE. A year later, Huskey was at UCLA's Institute for Numerical Analysis pioneering SWAC or Standards Western Automatic Computer, the newer brother of the National Bureau of Standard's SEAC. Kite Sharpless got command of the EDVAC project. Von Neumann and company churned toward completion of the Institute for Advanced Study's electronic machine. David Rees, a veteran of the top-secret wartime Colossus project at Bletchley Park, and I. J. Good, a Moore School lecturer, returned to England to sketch out and build the electronic Small-Scale Experimental Machine or SSEM, known at Manchester University as the "Baby." Mauchly and Eckert's very next offering was BINAC—or Binary Automatic Computer—for the Northrop Aircraft Company. There from the atomic energy facility in Oak Ridge, Tennessee, was Cuthbert Hurd, soon-to-be leader of IBM's Applied Science Department, from which he pushed the reluctant calculator company into the world of electronic computing.

Nestled in among the attendees was 30-year-old Claude Shannon, a Bell Lab electrical engineer and mathematician, who as an MIT graduate student had worked with Vannevar Bush and his Differential Analyzer. At the lectures, he was already famous for his MIT masters thesis, which he wrote in 1937. Entitled "A Symbolic Analysis of Relay and Switching Circuits," it was central to all of electronic computing by pointing out the identity between the symbolic logic and the binary values 1 and 0 of electronic circuits. By 1948, Shannon would become the founding father of information theory and the age of electronic communications. His paper, "A Mathematical Theory of Communication," based on his discovery of the similarity between Boolean algebra and telephone switching circuits, became the basic theory that explained the communication of information.

If Ben Franklin had been kicking around those lectures, he probably would have convinced the lot of them to throw in with the University of Pennsylvania. With the war over and everyone scattering for home and wondering what they would find when they got there, it might well have been easy pickings to convince many to settle down at the Moore School and to pioneer the future of electronic digital computing. Everything needed for a gigantic leap forward was just waiting for ignition. And they would not have needed Old Ben to finesse the federal government into putting up the cash. The Army was already there with an open wallet, and the economic belt-tightening but insightful and bright Harry Truman might well have agreed to go along with it all. Ah, so much for hindsight. By August 31st, *The Theory and Techniques for the Design of Electronic Digital Computers* was over, and its lecturers and attendees still had a lot of convincing to do to get everyone over to their way of thinking.

Maurice Wilkes saw Eckert as the lightning rod of the lecture series. "People who change the world," he said of Eckert, "stand facing two ways in history. Half their work is seen as the culmination of efforts of the past; the other half sets a new direction for the future."[53] Much the same can be said of the lectures themselves; a new direction was indelibly stamped on everyone there. They came as individuals and were galvanized into an electronic brotherhood. Everyone went home, quickly built machines, quickly spread the word, and quickly pushed electronic digital computing into the light of day. Issuing forth from laboratories both in the United States and England, no fewer than five new electronic, digital computers, the sons of ENIAC, powered up for all the world to see.

Mauchly and Eckert would produce the first commercially available wonder machine, UNIVAC, which would make a guest appearance on Edward R. Murrow's TV show, go on to predict Eisenhower's victory over Adlai Stevenson, and quickly become the synonym for computer in every household in America. Forrester and Everett understood their part in it all; they had to return to Massachusetts to design and manage the coming out of their electronic digital machine, to cut it in the direction of real-time computing, to keep it far from the grasp of mathematicians seeking only to crunch numbers, and to keep any hint of divisiveness at bay. Beyond that there was not much of a plan, but that was more than enough of a beginning.

Chapter Four
Escaping Old Ideas

*The difficulty lies, not in the new ideas, but in
escaping the old ones.*

—John Maynard Keynes

Like many natural disasters, man-made disasters can also
have aftershocks. The disaster of World War II would produce
many in 1946, most of which made their way to the Pentagon,
reverberating large and small throughout the corridors of the
massive structure. As with all aftershocks, there are origins
to be found out: who created them and why. Although the
sources might be readily pinpointed, the tricky questions
are where exactly they will go, how fast, and what will be the
extent of the damage when they arrive. The answers to the
aftershocks of World War II were anyone's guess. However,
if the brutal war had taught any lessons at all, an important
one was that a little foreknowledge about an adversary or
potential adversary might come in quite handy in trying to
avert the next man-made disaster. The aftershocks from Pearl
Harbor continued on for four years. But eventually the war
was over, the boys came home, and getting back to normal
was the country's only concern.

Getting back to normal would bypass the Pentagon completely and beating swords back into ploughshares would have
to wait. The US military, which had only a meager presence in
pre-1939 world events, had dramatically transformed into the
most powerful military in history by 1945, but by 1946 was
reverting to prewar levels as Truman "under intense public
pressure to bring the troops home and restore a sense of normalcy, ordered immediate demobilization and the withdrawal of almost all American soldiers overseas."[1] By December
of 1946 demobilization was nearly complete. "The pace was

*Cambridge Field Station (later, Air Force Cambridge Research Laboratory) 224 and 230,
Albany Street, Cambridge, MA, 1948 (around the corner from the Barta Building).
(Photo courtesy of USAF, Hanscom AFB.)*

so rapid that Truman would later claim that it was not demobilization, 'it was disintegration.'"[2] World tensions, however, were not demobilizing; they seemed to be mobilizing everywhere. As a result, the Pentagon's post-war halls were a flurry of activity, as it tried to catch up with every world event as it happened or, hopefully, before each occurred.

What was known within the Pentagon in 1946 wasn't good. The US military knew that America's great army of liberation was an army of occupation in Germany, Austria, and Japan, effectively reduced to a policing function "stripped of most armor and air power and assigned only maintenance of public order and denazification duties."[3] The rest of America's army was back home in "civies" waiting on the GI Bill. Large forces of British and French troops were nonexistent. "In the chaotic first two years after World War II, the pressing question on the minds of most Western European policymakers was how to survive the economic collapse the war had created, rather than how to maintain or build effective fighting forces."[4]

The Pentagon was well aware that a Soviet threat was beginning to rear itself up everywhere on the Continent, that George Marshall's attempt to relieve tensions between Nationalists and Communists in China would not last, that Ho Chi Minh's Viet-Minh would contest France's move to secure Cochin China, that Indonesia was all but lost to the British and Dutch, that the All-India Congress and the Moslem League could not share India for much longer, and that the Trans Jordan through to Iran was a powder keg of fomenting coups and revolutions. Everyone could see that Europe was in ruins. Greece was engulfed in a civil war. There was a new map for Eastern Europe that sported a bunch of new countries, and the old countries that were already there sported new governments. Both old and new in the East cowered under the threat of Joe Stalin's Red Star, which drew from Winston Churchill the famous remark that, "From Stettin in the Baltic to Trieste in the Adriatic, an iron curtain has descended across the Continent." On the heels of World War II, the Cold War had rumbled in quickly and in earnest. Where its tremors would lead was up to the enterprising folks in government and military intelligence to find out.

The Pentagon also knew that rapidly acquiring strategic information and then rapidly analyzing it were inescapable facts of life. World War II had driven that home with great force. The military learned quickly; and strategic and tactical information gathering became a key element to Allied success throughout the war. Radar information provided tactical warning about approaching enemy aircraft, ships, and armies. Radar combined with mechanical computers and servos enabled the targeting of weapons for devastating effect. Computational information was necessary in developing the atomic bomb. And cryptologic information derived from high-speed calculating machinery had cracked both Japanese and German secret codes and, as such, was instrumental in hastening the end of the war. All of that newfound expertise and more would be needed to sort out and prepare for any and all post-war aftershocks.

Ongoing from 1943 to 1946 was the effort by the United States and Britain to break the double-encrypted, Soviet diplomatic and KGB codes (the project was known as VENONA in the United States).[5] William Crowell, later Deputy Director of the National Security Agency, recalls how "gradually a picture of a massive Soviet espionage effort began to emerge from the work of the VENONA team,"[6] which would uncover the likes of Julius and Ethel Rosenberg, Klaus Fuchs, and Theodore Alvin Hall. The electronic digital computer was newly arrived on the scene in the form of ENIAC, and it might well be a good fit for assaulting stubborn codes or somehow helping to control the enormous influx of information. Having already shown its mettle in gun-firing calculations and atomic research, ENIAC—or an ENIAC-like machine—might well do the same for information gathering and analysis. ENIAC's reputation was well known in the Pentagon, and the Moore School lectures were high on its to-do list. Two top-secret military intelligence groups sent representatives to the Whistle Factory. Military code breakers from the OP-20-G (Office of the Chief of Naval Operations, 20th Division of the Office of Naval Communications, G Section) and the Army Security Agency were in attendance. The OP-20-G's James Pendergrass returned as an electronic convert and vigorously pushed his Moore School lessons on the Navy.[7] Both the Army and Navy were well aware that if the military presence abroad was reduced to a police force, then an army of eyes and ears was necessary to gather intelligence in its place. As the saying goes, to be forewarned is to be forearmed.

As the war drew to a close, continued access to ever-newer technology in peacetime was an ever-increasing concern to the military. In February of 1945, a month before President Roosevelt died, Secretary of the Navy James Forrestal, realizing that the Office of Strategic Research and Development (OSRD) was planning to close down at war's end, sent a note of concern to the White House. To Forrestal, scientific continuity and momentum were in jeopardy without the OSRD in operation. He sought direction on "how to establish channels through which scientists can [contribute to the nation's security by carrying on research] in peace as successfully as they have during the war."[8] Vannevar Bush, wartime head of the OSRD, replied in July of 1945 to then President Truman with *Science: The Endless Frontier*. "The Bush report recommended that a single new agency the National Research Foundation should be established to provide all such support, including defense and medically related research."[9] Although

Senator Warren Magnuson two weeks later introduced a bill in Congress to create the National Research Foundation, it was vetoed by Truman and languished until 1950, when it was finally realized as the National Science Foundation.

The veto came because the bill gave too much autonomy to the proposed national research organization, which was Vannevar Bush's rather naïve idea on how the country's research should be conducted, with scant accountability to the federal government. An old hand in the quagmire that is Washington politics as well as a frequent witness before innumerable Congressional committees, Bush might have been less bold in pushing his ideas on how the government should approach vital research. Bureau of the Budget director Homer Smith had testified before a Senate committee that, "an agency which is to control the spending of public funds in a great national program must be part of the regular machinery of government."[10] A five-year battle ensued between opposing parties. Slipping into existence during the fray was the Office of Naval Research (ONR), approved by Congress in 1946. Bush and his supporters went along with Congress on the ONR plan with the stipulation that "it was a temporary program"[11] to be funded only until Congress approved a national science and research organization. That would prove itself to be another naïve thought. Kenneth Flamm's *Creating the Computer: Government, Industry, and High Technology* shows how far from *temporary* the Office of Naval Research became over an extremely short period of time and how powerfully pervasive it would become in electronic computing:

> By the end of 1948, the ONR employed one thousand in-house scientists, funded about 40 percent of basic research in the United States, and was working on research contracts amounting to $43 million ($20 million of its own money, $9 million from other federal agencies, and $14 million of university money). Until the early 1950s, the only other government agency that exerted a role over the external technological development of computers even remotely approaching that of ONR was the National Bureau of Standards, and its involvement had a considerably different orientation.[12]

So much for *temporary.* Moreover, from 1946 to 1950, the Office of Naval Research funded a great number and variety of computer projects, including Whirlwind; Raytheon's Project Hurricane Computer (later renamed RAYDAC, for Raytheon Digital Automatic Computer); the CALDIC computer at the University of California, Berkeley; and the Harvard Mark III. These varied computer projects were put under the control of the ONR's Mathematics Branch, which meant that, like ENIAC, they would fall under the sway of mathematicians, with their emphasis on pure and applied mathematics, statistics, and numerical analysis. None of this augured well for the bright boys and their newfangled, all electronic, real-time, general-purpose, digital machine.

Navy code breaking was an ultra secret, furiously important activity and, therefore, immune from any influence from the computational frenzy of the Mathematics Branch. Although the Navy outspent the Army on high-speed code breaking and cryptoanalysis machines by a five to one margin beginning in the 1930s,[13] it was the Army with the decided lead in electronic computing. The Army owned ENIAC and EDVAC, and it shared with Navy Ordnance the funding for von Neumann's IAS computer. Electronic computing represented the future in code breaking and Naval Intelligence was quick to realize this and quick to react; however it could not wait, as the Army had done, three to five years to develop its own electronic machine. Besides, the existing machines were too out in the open; secrecy was paramount for CO-20-G's activities—secrecy not only from the Soviets but also from Washington politics "where the National Bureau of Standards was trying to gain control of anything called a computer."[14] The Navy first tried to retain its wartime cryptoanalysis crew, the top secret CSAW (Communications Supplementary Activities-Washington) group as peacetime civil service appointments. They all refused, preferring instead to return to civilian life. National Cash Register (NCR), the Navy's partner in providing high-speed analytical machinery during the war, also refused a Navy overture, opting instead to return to its lucrative office equipment business.[15] Captain Joseph Wenger, head of the Navy's cryptoanalytic section, convinced CSAW's Howard Engstrom, a Yale mathematics professor before the war, and William Norris, to form a private company and to do contract cryptologic work for the Navy. Ralph Meader, the wartime head of the secret Naval Computing Machinery Laboratory (NCML) in Dayton, Ohio, soon joined them. Private investment in the company seemed impossible to obtain after 17 companies refused to finance the venture, including the New York investment firm of Kuhn and Loeb. A savior was found when Meader located investment banker and retired Navy officer John Parker, who directed a glider manufacturing company, Northwest Aeronautical Corporation, during the war.

In August of 1946, Parker's former glider factory in St. Paul, Minnesota became the home of the new venture, Engineering Research Associates (ERA), with 40 staff drawn from CSAW, and others from the NCML in Dayton, Naval Ordnance, and the newly formed Office of Naval Research.[16] Parker was made ERA's president with Norris assuming the role of chief operating officer. Bending the rules for the young company, the Navy issued it noncompetitive contracts, plumping up its first-year revenues to $1.5 million.[17] ERA also got Navy help on the technical side of things in the form of captured German technology, especially that of magnetic drums.[18] Magnetic drum technology would become an ERA specialty, incorporated into many of its cryptological devices for the Navy, as well as machines for the commercial marketplace. In short order, another gush of technology would flow in ERA's direction when "the Navy gave ERA engineers access to reports on computer work on other government projects."[19] It represented a huge haul, since the Navy supported most computer projects going on in the United States.

In 1947, when the Navy ordered ERA to construct the Atlas computer, "it drew heavily on the work of others—logical designs produced by von Neumann's IAS computer group at IAS and engineering designs from MIT's Whirlwind computer group."[20] Well before Forrester and Everett had even designed their machine, they would be caught in a pincer movement of competing interests. On one side were the mathematicians and their penchant for quickly factoring the previously impossible, while on the other side were Navy initiatives like the Atlas computer seeking to grab off any new technological developments hot off the bright boys' or any other Navy project's blueprints. The rub came when Truman-era cost cutting sliced toward the ONR's Mathematics Branch, effecting its decisions as to which computer projects to maintain and which to jettison.

In 1946, the Army Air Force was not overly concerned about its lack of computing power. If you couldn't get the behemoth machines airborne, why bother? And if one could, what then—crack equations like the mathematicians were doing or decode Soviet message traffic? Computers, for the most part, didn't seem to have a place in the Air Force mission, but they bought one anyway. Mauchly and Eckert built the BINAC as an airborne computer ostensibly to be flown aboard the Air Force's Snark long-range missile, under construction at Northrop Aircraft. Somehow it was to work itself into the missile's celestial guidance system. However, as engineer Richard Sprague recalls, the computer never worked, and besides "the BINAC occupied the space of a small house."[21]

The Army Air Force seemed much more inclined to pay most of its attention to building jet aircraft and to forge on with its ongoing attempts to wiggle free from the Regular Army. Huge results of that effort were the F-86 Sabre jet and the F-84 Thunderjet, which were ready when war broke out in Korea. Hap Arnold, a five-star general (the only others were Eisenhower, Bradley, MacArthur, and Marshall), and a cohort of Air Force officers, including Generals Carl "Tooey" Spaatz (Arnold's successor to command the Air Force) and Lauris Norstad as well as Senator W. Stuart Symington, worked vigorously for an autonomous Air Force, but were a lonely bunch in the Pentagon of early 1946. There weren't many sympathizers to their plight. Army people did not want to lose direct control of their means to air-to-ground tactical support. Navy people did not want a threat to their air wing from an organization that styled itself as the nation's sole air power specialists. And the Marine Corps didn't want to get swallowed up into the Army when and if the big shuffle of services was brought before Congress to decide.

Luckily, the Army Air Force had friends in high places. President Truman backed it, as well as a general reorganization of the military from a Department of War to a Department of the General Defense with three co-equal branches, all answering to a civilian Secretary of the General Defense. General Eisenhower also backed Arnold's insistence on autonomy. He had seen what the Air Force was capable of during the war (he testified before Congress that Normandy would have been a disaster without the Air Force's mastery of the skies), and he was a firm believer in what he called "a three-legged stool" of coequal military organizations.[22] Once Arnold and company satisfied Eisenhower's demand that the new service would always have a tactical branch to support ground forces (the Army Air Force of 1946 was carved into three distinct branches: Tactical Air Command (TAC), Strategic Air Command (SAC), and Air Defense Command (ADC), Ike put his full weight behind Arnold's race for an independent Air Force. Truman consented to the Navy retaining its air arm and to the Marines staying put within the US Navy. With those deals struck, it was only a matter of time. And it would come to pass the very next year on the 18th of September with Spaatz as the first Chief of the United States Air Force and Symington the first Secretary of the Air Force. "We feel with deep conviction," said the new Secretary, "that the destiny of the United States rests on the continued development of our Air Force."[23]

German Plunder to the Rescue

The *continued development* that Symington called for would be a hard nut to crack. The Air Force didn't have a 1,000-technologist ONR with a stranglehold on national research, and it didn't have an Army ENIAC or EDVAC research background to maneuver itself into the coming future of electronic digital computers. Other than a few smallish research laboratories scattered around the country, the Air Force had only Wright Field in Dayton, Ohio as its central research facility. "The Air Force didn't have any technical support of its own," reported Ivan Getting, in an interview from 1995. "Other than Wright Field, the Air Force was entirely dependent on

General Arnold (left) and Theodore von Karman (right), circa 1945.
(Photo courtesy of USAF Historical Records Department.)

the Army Signal Corps for communications and radar . . . and on Army Ordnance for machine guns and ammunition and bombs."[24]

If the Soviet Union chose to attack Europe, it had over 190 divisions (with 10,000 troops per division) and over 10,000 aircraft versus an American force of 20,000.[25] A nuclear counteroffensive by US forces would have come to naught because the US did not have any atomic bombs. "David Lilienthal, chairman of the Atomic Energy Commission (AEC), inspected the atomic laboratory at Los Alamos in January 1947, he found only one atomic bomb that was 'probably operable.'"[26] Without the threat of bombers stuffed full with atomic bombs, the Strategic Air Command (SAC), formed in March of 1946, was rendered very unstrategic. Secretary Symington's *continued development* would have to be a homegrown Air Force effort and, hopefully, a mighty fast one. Although it did not have an ONR or ENIAC, the Air Force had something very powerful; it had a plan.

The plan came in 12 volumes replete with its own prolegomenon and summary. It was both a stratagem and a workingman's handbook that looked decades ahead into the technological future of the Air Force. Begun in April of 1945, a draft was rushed to completion and delivered into Hap Arnold's hands by August. The technical study represented the first time in US military history that such a highly detailed, future-looking plan had been devised.[27] Today such reports, usually far less than 12 volumes, are commonplace, expected, and necessary aspects of high-tech research and go by the prevailing moniker of *technology roadmaps*. However, this report was quite unlike any other plan to come before it or after it. For Hap Arnold, it bordered on an epiphany: in his hands was the key to America's modern Air Force and his long sought after wish list for aeronautical supremacy and air power . . . all courtesy of the German *Luftwaffe*.

In fact, the original title of the typewritten manuscript that Arnold first received was, *Where We Stand: First Report to General of the Army H.H. Arnold on Long Range Research Problems of the Air Forces with a Review of German Plans and Developments, 22 August 1945*. For the final, 12-volume set of December 15th, it bore a new title, and any German attribution was gone—*Toward New Horizons: First Report to General of the Army H.H. Arnold, Submitted on behalf of the A.A.A. Scientific Advisory Group*. Its summary volume, *Science: The Key to Air Supremacy*, introduced the extensive study and set the tone and direction for success. Recovered and on his way to meet President Truman at the Potsdam Conference, Arnold met up with his old friend and comrade, the aerodynamicist Theodore von Karman, in Paris on July 13th to get an update on Operation LUSTY.[28] The final composition, *Where We Stand,* would result from that meeting.

The acronym LUSTY was for the very top secret Operation LUSTY and was derived from: LUftwaffe Secret TechnologY. It was a mission conceived by the US Army for the exploitation of the *Luftwaffe's* secret aeronautical technology and took place directly on the heels of Normandy, beginning on June 6, 1944, in full swing by the 19th, extending past Germany's capitulation of May 8, 1945, and on into early summer. Operation LUSTY was a mad dash by the United States Army, Army Air Force, and Navy plus British Commandos with Canadian troops, and Free French forces—all operating independently of one another and, most times, in competition with each other—to grab up anything and everything they could of the secret workings of Germany's air force and rocket programs. Involving thousands of troops, specially trained for their sprint through the battlegrounds of Europe and carrying select shopping lists, called "Black Lists," of what to look for, Operation LUSTY was highly organized plunder on a monumental scale.[29] For the Air Force, it all began with a secret meeting in the backseat of a military staff car between General Arnold and von Karman, who together would soon thereafter be hailed as the principle architects of America's modern Air Force.

Eight months before Operation LUSTY, in September of 1944, as the war was grinding into its gruesome endgame, Arnold commissioned a plan to be set into motion that represented the Air Force's part in the Army's overall mission. In a dark, empty corner of New York's LaGuardia Airport, he rendezvoused with von Karman, his old friend and confidant from when Arnold commanded March Field near Los Angeles and von Karman was the director of the Guggenheim Aeronautical Laboratory at the California Institute of Technology (Caltech). Von Karman, who like John von Neumann was a Hungarian émigré and like his fellow country-man also carried a well-known, international reputation but in the field of aeronautics, was recognized as second only in the field to Ludwig Prandtl, his mentor during his student days in Göttingen, Germany.

As far back as 1912, von Karman had been director of Germany's Aerodynamics Institute in Aachen, and at the outbreak of World War I in Europe in 1914, he headed the Austro-Hungarian Aviation Corps.[30] "He trained many of the world's leaders in aeronautics, including the director of research for the *Luftwaffe* . . . [a job which] the head of the German Air Force, Hermann Göring, had offered . . . first to von Karman after he was already at Caltech."[31] And for Japan, von Karman had helped plan its first aerodynamics' institute in Kobe in 1928. Arnold had chosen his partner well. Better yet, Arnold and von Karman both had "a similar vision for military aviation: the United States needed a cooperative aeronautics es-tablishment that coupled civilian scientific and industrial expertise with the practical needs of the Army Air Corps."[32] Operation LUSTY provided Arnold with the supreme opportunity for a post-war Air Force to leap far forward in aeronautical technology, and provided von Karman with an equal chance to solidify Caltech as the nation's premiere aeronautical re-source and the West Coast as its foremost aircraft producer. As Arnold ex-plained to von Karman that night, "I want you to come to the Pentagon and gather a group of scientists who will work out a blueprint for air research for the next 20, 30, perhaps 50 years."[33]

This was Arnold's chance to lay aside the robe of conservatism that had weighed him down since his first meeting with Roosevelt in 1938 as chief of the Air Corps. In hindsight, with the outcome decades past and well estab-lished in history books, it is easy to minimize the enormity of that burden. Back then, it was anything but; it was the most dire of circumstances imag-inable. And it was all placed upon a single individual with no time for on-the-job training or a rookie mistake or two. Nothing in America's military history quite compares with it, except maybe Lincoln's entreaty to Ulysses S. Grant to help save the Union. The Navy's Battle of Midway, a brief five months after Pearl Harbor, was of tremendous significance both militarily and psychologically, but if it had been fought to a draw with the Imperial Navy or even if the US Navy had been beaten back, there was still time and the vastness of the Pacific to somehow recover. But to lose England, ah, that would have been the greatest of calamities. And the margin for error was ever so close.

In less than a year's time from Arnold's meeting with Roosevelt, Churchill was already giving last-stand speeches as with his signature address: "We shall defend our island whatever the cost may be; we shall fight on beaches, landing grounds, in fields, in streets and on the hills." Churchill's concluding line from that speech: " . . . until, in God's good time, the New World, with all its power and might, steps forth to the rescue and the liberation of the old," was meant for America's ears. That New World to the rescue was Arnold's Air Corps and America's factories churning out war armaments for the Royal Air Force. The burden of riding to the aid of England was Arnold's preoccupation from the very moment that he took command of the Air Corps. The paltry prewar output of America's aircraft and armaments had not only to be reversed but, in most instances, begun from scratch. The air power visionary in him had to take a backseat to the more pressing and practical business of filling convoys with vital war materiel for the battleground that was Britain.

Before the war, he had constantly and consistently preached research and development as the backbone of air supremacy. As a young Air Corps officer in World War I, he had grimly experienced firsthand the collapse of the country's preparedness. He watched as American flyers flew anything but American-built aircraft over the trenches of Europe, although the Wright brothers had invented heavier-than-air flight in 1903. Thereafter, he lived for pushing the envelope of flight ever further and cringed at "can't do" attitudes and excuses. Such a determined conviction was in particular evidence with his super secret building of the jet fighter aircraft, the Bell XP-59A, which lifted into the California sky less than a year after Pearl Harbor.

Arnold was not a newcomer to enthusiasm over aeronautics. It was a life-long passion. In 1911, when the 25-year-old signed up for flight school, his commanding officer encouraged him with: "Young man, I know of no better way for a person to commit suicide." He earned his military aviation rating less than a year later. Even later, in 1928, the year after Lindberg and his *Spirit of St. Louis* electrified imaginations with his Atlantic crossing, Arnold, then a major and writing as H. H. Arnold, penned the Aviator

Series of books for young boys, a six-volume set recounting the perilous adventures of a young airplane pilot named Bill Bruce. Published by the A. L. Burt Company, Arnold's books enjoyed a very popular following of youthful imaginations dreaming of emulating *Bill Bruce Becomes an Ace* or maybe becoming the next Lucky Lindy.

Years later, while in command of March Field, 40 miles from Caltech, he was but a quick drive from von Karman as well as his old buddy from the Great War, Robert Millikan, the Nobel laureate in physics in 1923, and at that time president of Caltech. Many an evening until the next dawn, Arnold had spent in Millikan's kitchen chatting up the high frontier. He was there from early on when Bill Boeing, Larry Bell, Donald Douglas, and Jack Northrop, the patriarchs of the West Coast aircraft industry, were first stamping their names on bold, new aircraft. He exulted in the rarified air of flashy metal monoplanes, radial engines, and young men doing steep dives over the desert. He was born to be a part of all the action. Ten years on, then a general, Arnold was still proselytizing the cause; his address to the 1937 meeting of the Western Aviation Planning Conference exhorted his audience always to, "Remember that the seed comes first; if you are to reap a harvest of aeronautical development, you must plant the seed called experimental research. Install aeronautical branches in your universities; encourage your young men to take up aeronautical engineering."[34] His ongoing mantra was "the first essential in air power is preeminence in research." And he lived it. He knew that was what would keep the Air Force first.

Soon after taking command of the Air Corps and before the heavy responsibilities of armaments diverted his attention, he took a last crack at doling out research assignments at a 1939 meeting of the National Academy of Sciences' (NAS) Committee on Air Corps Research. Millikan and von Karman were there and received marching orders to develop jet-assisted takeoff, which later became known as the JATO project. However, not every engineer or scientist there saw the direction of aeronautics Arnold's way. Take Vannevar Bush, for instance, who was there with Jerome Hunsaker from MIT. Bush was dubious of JATO. As von Karman later wrote in his autobiographical work, *The Wind and Beyond*, Bush could not fathom why "a serious engineer or scientist could play around with rockets."[35] As great an engineer and visionary as he was, Bush, like everyone else, had his blind spots. Three years later, however, in May of 1942, Bush and the Joint Chiefs of Staff created the Committee on Guided Missiles of the Joint Research and Development Board. Early on, he had also dismissed wasting government money on electronic computers like ENIAC, preferring that the country soldier on with mechanical and electromechanical machines.[36]

Hunsaker referred to JATO as "that *Buck Rogers* job." Arnold didn't appreciate the famous engineers' attitude; however, he had also fallen under the sway of bomber fanaticism and was late in coming to his senses about self-defending B-17s. So too would Vannevar Bush recover in time, see the light, and wholeheartedly trumpet rockets, missiles, and digital computers. The *Buck Rogers* job, unknown to anyone at that 1939 meeting, would become today's Jet Propulsion Laboratory of 5,000 employees with its multibillion dollar budgets and its handiwork seemingly rocketing all over the cosmos.[37] Aeronautics needed a champion and Hap Arnold was the right guy to take on the challenge. General Bernard Schriever, father of the Air Force's missile program in the 1960s, said of him, "There's no question, his greatness was that he created the infrastructure . . . visualized the kind of infrastructure that the Air Force needed to really get into the technology age."[38]

Rescuing England, however, had put a big crimp in that prewar zest. But, in the backseat of a staff car with von Karman, everything was all suddenly rekindled. Set before them was an unprecedented opportunity to take Air Force technology—that of aeronautics, rocketry, and electronics—and combine it with that of their soon-to-be vanquished foe. Certainly, they must have quickly and happily realized that such a magnificent whole would be years, maybe decades, ahead of its separate parts. They had already seen the deadly results of German technology at work for the *Luftwaffe*. Why not grab off the best of it for themselves? The war was edging to a close and German technology was there for the taking. Arnold wanted to be first, and like a couple of marauding pirates, they conspired that night to fill many chests with German booty. But they would have to be swift about it. For not only was there competition from the other Allies similarly bent on easy plunder, but there was also the Soviet Army hurtling toward the Rhineland and closing fast.

"Now to the Brocken the witches ride," wrote Goethe in *Faust*, about the famous peak set amidst the surrounding, mist-laden mountains and thick forests of the Harz Mountains, southwest of Berlin. Called simply the Harz by Germans, the mountains are the mythical gathering place for the traditional witches' sabbath of Saint Walpurga on the evening of every April 30th until dawn the next day. On *Walpurgisnacht* 1945, there were more than witches riding to the Brocken; Allied troops were racing toward the Harz set on techno-plunder.

Atop the Brocken, the world's first TV tower—which had broadcast the 1936 Berlin Olympics and daily aired the program, *Deutsche Reichspost*—was bombed on April 17, 1945 in advance of the military action. Not far away on the Elbe River, US and Soviet troops would meet up on April 25th.

The ominous rendezvous would set a clock into motion. For at the Yalta Conference in February, the Allies had divided the crumbling Nazi Reich into three zones of future occupation, with the Russians set to receive as part of their zone the very Harz where von Karman's marauders would plunder in search of the grist for Arnold's master plan. The Harz was well acquainted with plunder: the German word for it, *plündern*, was popularized there during the Thirty Years War (1618–1648) when thousands of unpaid soldiers laid waste to most every village and town as compensation for their services.

Given the temporary rank of major, von Karman officially became an Army Air Force consultant on October 23, 1944, and selected as his deputy the well-respected head of the National Bureau of Standards, Hugh Dryden. By December 1st they established the Scientific Advisory Group (SAG) and by January of 1945 had recruited 31 handpicked scientists and engineers who then met in Washington to set the plan into motion. Arnold advised the group to "throw conservative thinking to the wind."[39] He needn't have suggested it; they were already well committed to letting their imaginations ramble. Von Karman's raiders were enthusiastic and ready to actually see up close the advanced German technology that they had long studied only from afar.

Guy Stever, like his compatriots, relished the opportunity, saying that, "until this von Karman mission, we had to piece the enemy's facts together. Now we had the advantage of actually talking to the German scientists and engineers, seeing their laboratories, and hearing them describe their total programs."[40] January also saw Arnold suffer a severe heart attack that forced his untimely withdrawal to Florida to recuperate. From February to April, the SAG continued its meetings to develop and finalize the general format for their ultimate report to Arnold. They arrived in Paris on May 1st and found the Army's Operation LUSTY in full swing. Three collection depots had already been set up at the Merseburg, Nuremberg, and Halle airfields for shipping the spoils back to the United States.

As they soon discovered, the Harz was a hidden trove of *Luftwaffe* secret technology. Deep beneath its rocky surface were honeycombs of massive excavations—networks of tunnels that stretched for miles underground with large open spaces, some of which were thousands of feet square, for subterranean factories, buried research laboratories, and test facilities. And all of it was artfully and successfully hidden from Allied air reconnaissance. Operation LUSTY uncovered over 500 sites for exploitation, including the massive underground site at Nordhausen, built by 20,000 slave laborers from the nearby Mittelbau-Dora concentration camps (10,000 of whom died in the effort). The facility had its own crematorium capable of handling upwards of a hundred corpses a day. It was at Nordhausen that the infamous V-1 and V-2 rockets that fell on London were manufactured, tested, and then shipped to launch sites near the English Channel.

Stever went with the von Karman group to the village of Volkenrode near Braunschweig in Lower Saxony to inspect the Hermann Göring Air Force Research Establishment, which consisted of approximately 56 buildings, all above ground but skillfully designed—by one of von Karman's former students—so as to be completely hidden from aerial observation. "The engineers and scientists at this remarkable place," recalled Stever in his *In War and Peace: My Life in Science and Technology,* "worked on transonic and supersonic aerodynamics, both theoretical and experimental, using wind tunnels for aircraft and firing ranges for artillery projectiles and rocket-propelled missiles."[41] George Schairer, Boeing's lead designer on the B-47 jet bomber project, noted a swept-back wing model in a wind tunnel, chatted up a few Germans about its significance, and then dashed off a telegram back home to sweep back the wings on the B-47. That one telegram gave Boeing a huge lead in building high-speed subsonic aircraft . . . while saving a fortune in research. Operation LUSTY's Air Technical Intelligence teams or "T-forces," operating under the Exploitation Division of the Directorate of Intelligence, discovered "rocket-propelled fighter planes, radio-controlled bombs, guided antiaircraft missiles, and practically every type of fighter aircraft in the *Luftwaffe* inventory."[42]

The techno-spoils went far beyond everyone's wildest dreams. In addition, more than 100,000 tons of documents were confiscated. Everything was quickly sifted through and evaluated by Combined Advanced Field Teams, with the real good stuff held for SAG members to examine. The plunder was then shipped either directly back to the United States or, as in the case of documents that were not critically important, forwarded to a specially created holding library in London.

Far to the north of the Harz, barely a day before the Soviet Army arrived at the missile launch site at Peenemunde on the Baltic, the United States Army had already absconded with the goods. Peenemunde launch-site equipment and Nordhausen V-2 stocks were shipped to Antwerp and then to New Orleans. Eventually, 300 boxcars of pillaged technology snaked along the New Mexico desert to a permanent rendezvous at White Sands. From Messerschmitt facilities in Augsburg and Obermmergau in the south to Volkenrode in the Harz, aircraft, rockets, missiles, and test equipment of every description, especially jet fighters, went to the United States as expeditiously as possible. Flown or carted to Cherbourg, the aircraft were stacked aboard the small attack carrier HMS *Reaper*, borrowed from the Royal Navy, and then shipped out on July 19, 1945 to New Jersey, where the cargo was selectively redistributed to either Wright Field, Ohio or to Freeman Field in Indiana. Overstock, and there was much, was stashed in C-47 hangers owned by Arnold's friend Donald Douglas of Douglas Aircraft at what is today O'Hare International Airport in Chicago.

The highly systematic techno-plundering quickly came to an end. The biggest advantage resulting from its methods was that it had been *intelligent* techno-plunder: enemy advanced technology finely combed by people who knew what they were doing, especially people who were themselves involved in very similar but somewhat less advanced research. Armies are good at learning from each other through warfare; the *knowledge transfer* from vanquished to victor has many times proven itself to be decisive. The Roman legions in conquering Spain quickly admired the benefits of the enemy's double-edged sword and seized it for themselves. Calling it *gladius hispaniensis* or sword from Spain, their new weapon soon became a decisive part of their world conquests.

Americans following World War I weren't as clever. Arnold had a long memory of it, and he was not about to allow his Air Force to be twice a fool. Arnold Pacey points out in his *Technology in World Civilization* that what is important is the intelligence behind the transfer. "Often, the 'transfer' of technical knowledge or equipment from one country to another has initiated a process of modification and adaptation from which new and inventive ideas have emerged. When these people are already developing related techniques, their dialogue with the new technology is likely to be especially creative."[43]

Some have argued that Air Force technology leapt ahead by five years at a saving of over $15 billion as a result of their techno-plunder. The spoils of the *Luftwaffe*, as evidenced in the B-47's wings getting a 35-degree backwards sweep, seem to agree. If even half true, Arnold's commissioning of the 12-volume study was a stroke of genius. From the actual technology brought back, to the plans, blueprints, even to the scribbled notes and journals that had led to the technology—all were minutely sifted, studied, and scrutinized by the collective intelligence, talent, and expertise of Arnold and von Karman's Scientific Advisory Group. Their analyses soon led to synthesis and then quickly to the typewriter. Von Karman and company put together a working document for Hap Arnold to read, consider, and act upon. Before returning home, von Karmen made one final stop together with a few close associates. They flew to Göttingen to visit the 70-year-old Ludwig Prandtl, ill and at home. Stever remembers that "the two giants could hardly speak and the rest of us could hardly guess what they felt."[44]

It could have all ended right then and there with the 12-volume *Toward New Horizons: First Report to General of the Army H.H. Arnold, Submitted on behalf of the A.A.A. Scientific Advisory Group* or simply known as *Toward New Horizons,* becoming an expensive coffee table set for Arnold's Pentagon office. He was not a healthy man (he had suffered multiple heart attacks by 1946), America was demobilizing back into a state of unreadiness, and the Air Force had virtually no research and development facilities to conduct the needed work. Carl Spaatz would take over on July 1st, and it would have been so easy for Arnold to walk quietly away into retirement as a five-star general and war hero. Book deals, movies, the lecture circuit, government consulting, and corporate boardrooms were his for the asking. However, Arnold, between December 15, 1945 and his retirement on June 30, 1946, chose to act, to implement, and to do so as fast as possible. War budgets, he reckoned and reckoned correctly, would be drastically slashed before too long. He needed to spend, and he did.

His problem was how to glean from the 12 volumes the best of the technology to institutionalize within the Air Force; and if he could pull that part of it off, then where would he turn for research and development? He was convinced that his Scientific Advisory Group must be made permanent. "These men did things that the average Army officer could never have accomplished," he told Spaatz, reminding him that the Air Force had no labs and no great scientists. "We must not lose these contacts."[45] On July 1st Spaatz did exactly that: the SAG was made permanent, and its name was changed to the Scientific Advisory Board (SAB) with von Karman running the show.

As to the *what-comes-first* part of the situation, von Karman provided the answers. In what he termed "fundamental realities" or where the future direction of the Air Force should be bound, he extracted from *Toward New Horizons* eight salient technologies needing serious attention:

1. Aircraft—manned or pilotless—will move with speeds far beyond the velocity of sound.

2. Due to improvements in aerodynamics, propulsion, and electronic control, unmanned devices will transport means of destruction to targets at distances up to several thousand miles.

3. Small amounts of explosive material will cause destruction over areas of several square miles.

4. Defense against present-day aircraft will be perfected by target-seeking missiles.

5. Only aircraft or missiles moving at extreme speeds will be able to penetrate enemy territory protected by such defenses.

6. A perfect communication system between fighter command and each individual aircraft will be established.

7. Location and observation of targets; takeoff, navigation, and landing of aircraft; and communication will be independent of visibility and weather.

8. Fully equipped airborne task forces will be enabled to strike at far-distant points and will be supplied by air.[46]

The eight were startling. Except for the technology listed as #3, "Small amounts of explosive material will cause destruction over areas of several square miles," which was Hiroshima and Nagasaki with a decided lean toward thermonuclear bombs, they were eye-openingly original. Neither Allied nor Axis air forces ever attempted anything like them during the war. Chennault, Saville, and the crews at the Chain Home operations center would have felt right at home with the offensive-defense perspective, the tight communications, and all-weather capabilities. Probably not so for the bomber cronies at Maxwell Field; von Karman's "fundamental realities" exposed the folly of their mania for high altitude, daylight, precision bombing (HADPB). Von Karman's research confirmed what every bomber crew over Europe knew: "This is crazy!" Von Karman and his SAG crew had used their expertise, what knowledge they had plundered during Operation LUSTY, and what their imaginations made of it all to create a third thing that was a remarkable scenario of what was to come. Hap Arnold's request for a 50-year look into the future of the Air Force was more than realized in *Toward New Horizons*. The 12 volumes were powerfully resilient and long lasting; they helped in winning the Cold War of 1945–1989, and were not replaced until 1995.

From the same pages von Karman also discovered some of the reasons for the success of *Luftwaffe* technology, reasons that the Air Force might emulate to its benefit. Three cropped up over and over again throughout the 12 volumes, and von Karman was quick to note them and to backtrack to the all-important sources from which they sprang. Consistently, he felt, the Germans had three things nearly always in their favor: an early start to their research, unlimited funding, and boldness of execution.[47] None of these things were by happenchance. An early start with plenty of military backing and plenty of money are understandable when you are the aggressor taking on the rest of the world. The real intangible was "boldness of execution." The *Luftwaffe* and its development system were well-oiled parts with decades of experience together. The Reich Air Ministry (*Reichsluftfahrtministerium* or RLM) was quite effective, with a few notable exceptions, at maintaining that essential balance. Yes, *Luftwaffe* research and development were bold; and sure, its execution was masterful. But why and how? America's Air Force would similarly have to boldly execute to be successful, and the blueprints for that human dimension were nowhere to be found in the warehouses full of *Luftwaffe* loot.

Such boldness was not a new phenomenon that sprung magically out of the Third Reich. Anold Pacey in *The Maze of Ingenuity* shows German self-confidence in technology ushering up for over three or four centuries. It is an ethos deeply rooted in Germanic history; a cultural and intellectual self-assuredness with an ancestral foreshadowing dating back to the chemistry of Paracelsus, the practical philosophy of Comenius, and the teaching of science from August Franke at the university of Halle.[48]

It was a centuries-old combination of applied science and practical craftwork; a certain intrinsic harmony between eye and hand that instinctively knows the molten metal in the crucible is ready and at the same time understands perfectly well how the interactions within the glowing liquid came to be that way. Much of that national character trait had its foundation forged centuries before in the same oft-plundered Harz, the resting place of the legendary Frederick Barbarossa and the glory of the Teutoni. It is probably no coincidence that great halls of learning like Halle and Leipzig are just to its east, Göttingen even closer on the west, Weimar and the Bauhaus a quick 50 miles to the south, and that it drew both von Karman and von Neumann to their studies at Göttingen.

Swiftly duplicating such technological prowess in the United States, even in the face of Cold War threats, seemed utterly impossible for the foreseeable future. With the Soviet Army hustling toward the Harz, the most expedient solution seemed to be to plunder people as well as technology. That's exactly what happened. The US Army initiated Operation Paperclip and immigrated some 750 German scientists and technologists into the United States (in some cases sanitizing questionable Nazi backgrounds). Werner von Braun, his brother Magnus, and most of the Peenemunde crew (including their boss Walter Dornberger), were *paperclipped* out of the Fatherland. Germany would have been unable to put them to work anyway; making military jets and rockets again were strictly *verboten*; and Germany's cities were craters of rubble with any economic recovery off in the distance. Soviet plunderers would have provided far less gracious living conditions in Russia for the *Luftwaffe* scientific braintrust. Operation Paperclip was an easy sell.

On the other side of the Atlantic, America's true genius was and is all about pioneering new frontiers. Thomas Hughes says of Americans in the Introduction to his *American Genesis: A History of the American Genius for Invention,* that their "most notable and character-forming achievement for almost three centuries has been to transform a wilderness into a building site."[49] Whether pioneering transcontinental railroads, electricity, oil fields, or space travel, Americans seem to have a decided knack for creeping their civilization over hill and dale. Starting with barely four million colonists hugging the east coast of the United States, America pioneered its way 3,000 miles to the Pacific. Such a clever show of native talent offered rewards as intangible and as powerful as those indigenous to the Harz.

Toward New Horizons was a wilderness document; it described a new frontier ready to be pioneered and transformed. Over time, and no one knew how much time, a totally new infrastructure—from basic education to the retooling of every major military and industrial laboratory—would need to take place in the United States in order to realize its goals. *Toward New Horizons* was not only good for the Air Force; one should have been written for the entire country as well. And 1946 would have been a favorable starting point; the country need not have waited ten years until Sputnik.

Von Karman also noted what the Germans had not done and that it was just as telling as what they had done, maybe more so. For the *Luftwaffe,* "most of the development took place in the industrial laboratories . . . but the very brilliant groups of German physicists in universities were never called in to participate. Consequently, while engineering design was good, imaginative new thinking was lacking."[50] Among established German laboratories, however, there was generally openness to new thinking within groups of colleagues, as with the spread of Adolf Busemann's seminal theory on wing sweep for supersonic flight. From it, both Messerschmitt and Focke-Wulf built swept-wing jet fighters. The United States, on the other hand, had a more open development system. Places like the Rad Lab and other OSRD contractors had their staff members but also a steady influx of talented graduate students—Bob Everett for one, coming aboard the Servomechanisms Laboratory after graduating from Duke in 1942.

Advancing technology that the United States was already good at—especially that of microwave radar—was another crucial part of *Toward New Horizons.* Von Karman felt that the "development and perfection of radar and the techniques for using it effectively are as important as the development of the jet-propelled plane."[51] Inklings to the advantages resulting from radar are everywhere in his eight "fundamental realities." So too are thoughts on airborne electronics—avionics well before the technology was created and the word coined. And between the lines of his "perfect communication system" in #6, and #7's "location and observation of targets; takeoff, navigation, and landing of aircraft; and communication" were the nascent glimmerings of a few electron tubes and the hum from an invisible world deep within a box.

Back Home

If there was a catbird seat of computing at MIT in 1946, it would have been a two seater for young Forrester and Everett. High in their perch, they could survey below all the geared gizmos, some pretending to be computers, that spread out over the campus. The Bush Differential Analyzers from the 1930s were still there, as was the Rockefeller Differential Analyzer (an attempt to make an electronic version of a Bush Analyzer), plus IBM 604 and 602 calculators, punched-card machinery, and some engineering plans from Wilcox Overbeck for another analog device to be called the Rapid Arithmetical Machine, which never got built.[52] Work on the Rapid Arithmetical Machine was halted in 1942. With their return from the Moore School that August, they were convinced more than ever that a remarkable opportunity had been set before them to build a very extraordinary type of computing machine.

The essential difference in their favor was that they intended to build a control computer—a machine that would automatically control the actions of yet another machine. Early on, that other machine was a flight control analyzer. Seemingly everyone else had a fixation on constructing single-purpose, scientific computers used solely to hurry answers to previously intractable mathematics problems into being. The scientific computer factors an algorithm's recipe of equations until it produces an answer. Replace the algorithm with yet another and it produces yet a new solution. Need to know the winter weather patterns over the Atlantic Ocean for the past 50 years? Feed an algorithm and a half-century of Atlantic winter weather data into a scientific computer, and it will generate an answer.

However, replace Whirlwind's flight analyzer/simulator with a different machine, like one to operate a skyscraper's lighting system or another to open the spillway at Hoover Dam, and the computer alters their reality: lights pop on and off in the skyscraper and millions of tons of water cascade out of a mammoth dam. Place Whirlwind into the practical hands of engineers surrounded by countless other machines, and the possibilities of linkage begin to blossom. Although in 1946 Forrester and Everett did not have an inclination toward controlling anything other than a flight simulator/analyzer, the skids were already greased and waiting. High above the gathering crowd of computer pioneers, they were alone with a gathering realization the implications of which just might turn out to be the perfect complement to *Toward New Horizons* . . . and a whole lot more.

The bright boys had no competition anywhere at MIT, or uptown at Harvard, or in Cambridge or Boston, or for that matter, anywhere south of the Charles River right to the doors of the Whistle Factory. Although they had yet to set out even the barest of plans to their electronic digital computer, they were far ahead of the outdated and embarrassingly analog machines around them. ENIAC's capabilities and its promise were just too profound to overcome. And of these clunking giants none would ever witness a Licklider penning a paper exulting the potential of man and *analog* machine symbiosis.

MIT's kingdom of gears was called the Center for Analysis and was presided over by Samuel Caldwell, colleague and partner with Vannevar Bush in the making of the Differential Analyzer. Although in the summer of 1945, the Navy offered to finance the construction of an electronic digital machine if the Center would submit a plan for one, Caldwell declined.[53] It was a good decision because neither MIT nor Caldwell had the capacity to pull off such a request. It was a little like asking a skilled sword maker to suddenly start producing ray guns. And although MIT vice president James Killian reported to the school's president, Karl Compton, that the Center was still a going concern and had "contacts, reserve funds, prestige, and staff" to be a key player,[54] it really did not have the foggiest idea of what was afoot in electronic digital computing. The only true shot at MIT producing such a machine was up in the catbird seat with the bright boys. "We were engaged in an endeavor," recalled Forrester, "that no one else could claim to understand."[55]

While Sam Caldwell listened to gears grinding to a halt around him, postwar aftershocks were rumbling beneath the venerable foundation of MIT's School of Engineering. The wartime work of the Rad Lab, dominated by physicists, had exposed the electrical engineering curriculum as being as backward as Caldwell's gear factory. Stuck in the mold of the classic electrical laboratory with a mindset, expertise, and tools for large-scale, electrical power engineering, MIT engineering was in dire needed of serious revamping toward electronics, electromagnetic theory, communications, and circuit theory.[56] Harold Hazen, dean of the department, said rather bitterly—while at the same time liberally spreading the blame around— that "Engineering over the country as a whole did not have the intellectual standing that science had."[57] Of course, intellectual standing had little to do with the situation. A dated curriculum has no intellect; it is just out of touch with reality. Then too, this was also a case of the boys having at each other again. It was the same mewling dogfight between engineers and physicists that went on during World War I when Thomas Edison refused the American Physical Society (physicists) and the National Academy of Science from participating in war research through his Naval Consulting Board. The reason given by Edison's chief engineer, M. R. Hutchinson, was that the Naval Consulting Board should have a membership "of practical men who are accustomed to doing things, and not talking about it." [58] The spat ended when George Ellery Hale, director of the Mount Wilson Observatory, eventually persuaded President Woodrow Wilson to establish the separate but equal National Research Council for scientific contributions to the war effort.

Hazen had good reason to be so defensive; physicists had blown the doors off the Rad Lab with pulse circuits, microwaves, feedback, and quantum mechanics. Their astounding displays made glaring the inadequacies of engineering. Karl Compton, as a physicist and president, well understood such situations. He had experienced two of them—once from his days as a Princeton physicist while reconstructing its physics department, and then again at MIT, when he hired John Slater, together with Philip Morse, Wayne Nottingham, and Robert van de Graaff, to elevate MIT physics. During the Slater-led transformation, Compton had eased out the old physics chairman and five senior faculty members to help the process of change run more smoothly.[59]

He was again faced with a similar challenge, this time in the field of electrical engineering. Engineers like Forrester, Everett, Brown, and the entire Servomechanisms Lab were already well indoctrinated into the wonders of electronics from their stints working Rad Lab projects. J. Presper Eckert had likewise learned about the new arts in roughly the same manner, having done wartime contract work for the Rad Lab at the Moore School; so too had William Shockley (co-inventor of the transistor in 1948) gotten similar exposure through his invention of the mercury delay line for Rad Lab radar. Electrical engineers without such exposure were, as Ivan Getting felt, way behind the eight ball. Sam Caldwell was one of them, and all he ever needed to get caught up was to catch the ear of a very able digeratus in his own Center for Analysis, one Perry Crawford, who had convinced Forrester to go electronic with Whirlwind.

Historian Larry Owens points out from correspondence between MIT's Julius Stratton and James Killian that it was common knowledge at MIT that the postwar Center for Analysis had lost the confidence of other departments and had begun visibly to crumble.[60] As John Maynard Keynes observed wryly, "The difficulty lies not in the new ideas, but in escaping the old ones." If the war had speedily transformed cloth-covered biplanes into metal monoplanes with swept-back wings and jet engines, then what the heck had Sam Caldwell been doing all the while? Was he under a rock or at the controls of a bygone era flying in circles?

Caldwell flirted with success in 1946 when MIT applied for and received a $100,000 grant from the Rockefeller Foundation, *not* to build a computer "but rather to study how to bring digital and electronic technologies into MIT's computing program."[61] In terms of 1946 dollars that was a helluva rich offer for a technology that the bright boys were already bringing to the university. But, who's to argue with a tidy and very timely grant for the Center for Analysis? The Center's lifeblood seemed to be grants; it was established with a grant from the Carnegie Corporation to Harold Hazen's electrical engineering department, then sustained by a Rockefeller grant to build an electronic version of the Bush Differential Analyzer. MIT returned part of the grant when the conversion faltered and was stopped.

But in 1946 Caldwell and MIT were back at the Rockefeller money trough for more, and got more. The carrot in the deal for the $100,000 was the intimation that electronic computation at the Center for Analysis might lure MIT's resident math genius, Norbert Wiener, to set up shop there.[62] It didn't happen; Wiener showed no interest and went off to Mexico to pursue the intellectual trajectory of his cybernetics research. Thereafter, Caldwell's time and that of the Center for Analysis grew short. MIT chose not to put any money or effort into sustaining the moribund facility. The Center and Caldwell sank quickly and disappeared from sight. It took a few years for MIT to swallow such a defeat, but in "1950 MIT Provost Julius Stratton confessed to then-President Killian that MIT had muffed its chances to establish leadership in computing."[63] Phoenix like, it arose again ten years later, in 1957, as the Center for Machine Computation with Phil Morse, a physicist, at its helm.

Postwar aftershocks also unsettled the way MIT financed its basic research operations. Going forward, Whirlwind was a good future look at the coming research model: military projects done with military money for military technology that may eventually in some manner, shape, or form go commercial. The prewar days of philanthropic gifts, grants, and industrial funding alone pulling the research wagon along were gone; and any thought of returning to them was an idealistic vision that would "turn out to be an endlessly receding mirage."[64] The world had big problems that required big science to fix, which needed real big money that only the government's wallet could cover. Whirlwind's benefactor, the newly hatched Office of Naval Research, took the lead in funding university science and electronics. "Bolstered by millions of dollars from recently cancelled procurement contracts, the ONR became the predominant patron of academic research, and set the precedent for postwar relationships between the military and the university."[65]

The Army was not far behind. A memo—more like a veiled command—from General Eisenhower, dated April 30, 1946, was given to Compton and Stratton:

> The lessons of the last war are clear. The armed forces could not have won the war alone. Scientists and businessmen contributed techniques and weapons which enabled us to outwit and overwhelm the enemy. Their understanding of the Army's needs made possible the highest degree of cooperation. This pattern of integration must be translated into a peacetime counterpart which will not merely familiarize the Army with the progress made in science and industry, but draw into our planning for national security all the civilian resources which can contribute to the defense of the country.[66]

The Air Force checkbook was open as well, with Hap Arnold and von Karman penning wish lists and setting them into action. At MIT such military precedent wasn't startlingly new; it had been firmly entrenched *during* the war. The university hoped it would all somehow wind down after V-J Day and then close out completely with the departure of the last of the Rad Lab staff in late December of 1945. It didn't and wouldn't. Ike articulated the military's needs in no uncertain terms: "techniques and weapons, which enabled us to outwit and overwhelm the enemy." The military wanted more of the same.

During the war, MIT was the largest recipient of research contracts with $56 million from the Office of Scientific Research and Development (OSRD), which accounted for 20% of the total doled out to all other educational institutions combined.[67] There was another, larger chunk of $1.5 billion in contracts that went to industrial firms that manufactured the fruits of the Rad Lab's labors. Companies like General Electric, Western Electric, Raytheon, Sperry, and Westinghouse had made huge, wartime fortunes. Profits soared especially in the new industry of making microwave tubes. "Reflecting on the boost the war and the Rad Lab had given to tube research, [the Rad Lab's Louis] Smullin joked, 'We were the real war profiteers; there's just no question about it.'"[68] "At the end of World War II, MIT was the nation's largest nonindustrial defense contractor, with seventy-five separate contracts worth $117 million, far ahead of second-place Caltech

($83 million) and third-place Harvard ($31 million)."[69] Flamm's *Creating the Computer: Government, Industry, and High Technology* points out how early on there were those quite peeved over Forrester and Everett's ONR cash. For some established members of academic mathematical circles, who sought out funding, it was especially galling "to be in close competition for research dollars with a group of young, largely unknown MIT engineers, with no real finished product to show for their efforts."[70] Engineers beware mathematicians.

When it comes to money, there's never enough, especially after four years on the wartime gravy train. But in postwar research, if a mathematician did not have a computer like the great mathematician von Neumann, he was considered undressed and, therefore, unfit for funding. "Would he were fatter!" lamented Shakespeare's Julius Caesar, eyeing the lean and hungry look of Cassius. Remembering the tumult of Mauchly and Eckert at the Moore School, Forrester and Everett were wise to exchange their catbird seat for the relative anonymity of the Servomechanisms Lab, all the while keeping an eye peeled, not for Roman togas, but for guys whose scabbards were filled with slide rules.

In the wake of the Rad Lab closing its doors, MIT and the military cemented a partnership together in March of 1946 with the founding of the Research Laboratory of Electronics (RLE). MIT's industrial pals were financial no-shows for the new lab; the same defense contractors that made huge profits from wartime contracts with the Rad Lab now offered in return to sponsor a few graduate fellowships. Truman's veto of the National Science Foundation Act (1947) squashed any help from Vannevar Bush and Congress. Philanthropies backed away at the scent of the military. Individual donors saw the Laboratory well cared for by its deep-pocketed, military patrons and either smiled in putting their billfolds away or went looking for others in greater need.

With a million dollars of government surplus equipment and $600,000 in government cash, the Research Laboratory of Electronics opened for business with 17 MIT faculty members, some ex-Rad Lab transfers (especially Jerome Wiesner, Albert Hill, and Jerold Zacharias) plus 27 staffers. Julius Stratton, RLE's boss, was ever the pragmatist, explaining that the arrangement "sets a pattern for the proper sort of relation between an academic institution and a sponsoring agency."[71] *Sponsoring agency* would become the new, elastic euphemism for military or military-related activities. Some openly feared the new funding arrangement. Harlow Shapley was one, saying "Those who were worried about domination of freedom in American science by the great industries, can now worry about domination by the military."[72]

Like an old uncle arriving with his suitcase for the weekend then staying forever, the military lingered on at RLE, accounting for all but about three percent of its budget throughout the 1950s.[73] In addition, RLE regularly ran conferences for the military on the latest developments in microwave communications, telemetering, and so on; its inaugural conference featured a keynote address by Zacharias, titled, "Basic Research—A Military Asset."[74]

More than a few of RLE's researchers would eventually slip off on their own, starting over a dozen companies, most of which specialized in microwave devices for military sponsors. Zacharias is a good example. His work on atomic clocks, for which all three services provided contracts, was transferred to a private manufacturer, the National Company, in nearby Malden, Massachusetts where Zacharias was its chief consultant. The military subsequently bought nearly all of the atomic clocks produced by the National Company, which later produced the first commercial version, the Atomichron.[75] The cesium atomic clock, a descendant of the Atomichron, is the engine of accuracy in today's Global Positioning System (GPS).

That success, in turn, provided Zacharias with a springboard in founding his own company, Hermes Electronics, together with Jerome Wiesner (RLE director, National Company board member and technical adviser, and eventually 13th president of MIT). To wit, the old adage that nothing succeeds like success. Julius Stratton would give the old uncle the key to the house when in January of 1946, on being named chairman of the government Guided Missile Committee, he wangled for RLE a multimillion-dollar contract for the Navy's top secret, air-to-air missile project called Meteor.[76] Then, with Guy Stever back from Operation LUSTY and looking to begin his career again, Stratton convinced him to become executive officer of Project Meteor.

Project Meteor was just the tip of the rocket/missile craze sweeping through the military. A Navy surface-to-air missile, Project Bumblebee, was in full swing at the Applied Physics Laboratory at Johns Hopkins University. The Army had German V-2s hurtling skyward at White Sands, New Mexico, plus another air-to-ground research project, Hermes, with General Electric, the Nike surface-to-air missile at Bell Labs, and the Army's long-running Corporal missile project. Hap Arnold had no less than 16 missile projects underway, including the long-range Snark with Northrop Aviation; Martin Aircraft's Matador cruise missile; and another long-range craft, the Navajo, with North American Aviation. Bill Boeing's bomber factory worked on the MX-606 surface-to-air supersonic missile, and Project Wizard had a home at the University of Michigan's Willow Run facility, which also ran the Guided Missile Training Program for junior and senior Air Force officers in guided missile technology. Hap Arnold's command staff, still bomber guys to the hilt, went along with Arnold's foray into missile technology. "At least partly as a result of *Toward New Horizons* . . . [Air Force planners] embraced von Karman's predictions on the feasibility of ballistic missiles and inserted a missile development program in the five-year R&D projections."[77]

Arnold pushed the edge of his techno agenda into a new incarnation of operations research: think tanks. When he needed a place where "experts could analyze and propose advanced concepts for the Air Force,"[78] he anted up $10 million for his good buddy, Donald Douglas (father-in-law to Hap's son), to spend on Project Rand (an acronym born from Douglas Research ANd Development). Project Rand later became the RAND Corporation—the Air Force's very own think tank and living extension to *Toward New Horizons.* Project Rand's first job for Arnold leapt the Air Force into outer space with its report: *Preliminary Design of an Experimental World-Circling Spaceship.* In May of 1946, Arnold struck again, this time with a $1.3 million feasibility study on *Nuclear Energy for the Propulsion of Aircraft,* known as NEPA. MIT called it "Project Lexington," and by 1951 the budget grew to $8.3 million. ("President Kennedy 13 years and $900 million later killed the program."[79]) Arnold's forays into jet fighter aircraft development continued unabated and in a hurry with the F-86 Sabre, F-84 Thunderjet, and the B-47 jet bomber. MIT's aeronautical engineering department tried to recruit von Karman away from Caltech; failing that, they lured away von Karman's best student, H. S. Tsien, another SAG veteran of Operation LUSTY, who figured prominently in helping author *Toward New Horizons.* MIT shared heartily in the largess of its sponsors' push to convert their plundered German technology into next-generation, All American air, land, and sea power. One opportunity that Arnold missed on his assault at cornering aeronautics for the Air Force was when the Navy funded MIT's supersonic wind tunnel, capable of operating at Mach 2 to Mach 4 (Mach 2 is generally twice the speed of sound; Mach speed is named for its Austrian inventor Ernst Mach).

Rockets, missiles, and jet planes were easy for Arnold to get excited about; the plundered technology was right there for him to see stacked up in hangars and warehouses stateside. He understood how close the Germans had been to going from subsonic to supersonic; it all just shimmered there in front of him with swept wings, sleek airframes, fins, and nose cones. He knew instinctively how to go after such a technology transfer, and felt even more confident with *Toward New Horizons* under his arm. More elusive was his feel for those technologies that von Karman had considered most important: radar for one and electronic navigation for another. If he poked around amid the German booty or searched through *paperclipped* researchers, he would uncover neither advanced radar technology nor the people who made it work, because it was an Allied specialty. So what does a wise man do when he really wants to know about something that he knows nothing about? He finds the people who do, and then hires them. As Arnold had told Carl Spaatz, brilliant scientists and engineers were in short supply within the Air Force, so he went looking.

Arnold turned to his best radar research facility: the Bedford Army Air Field, in Bedford, Massachusetts, a dozen miles from MIT. Renamed Hanscom Field in 1942 in honor of Laurence Hanscom, a news reporter who died in an air crash in the nearby town of Saugus, Hanscom became during the war the prime testing site for Rad Lab radar. The Bedford facility was run by John Marchetti, the same guy who at Fort Monmouth in 1937 demonstrated for Arnold and Saville his radar system for spotting bombers. Stowed away at the air field were advanced prototypes as well as the latest in operational radar units. Marchetti was ordered to set up a permanent research sub-station in Cambridge and to recruit Rad Lab, Harvard Radio Research Laboratory, and Boston University radio electronics and radar experts during the postwar exodus from their OSRD work. Marchetti's recruitment venture and the new Cambridge facility bore immediate fruit. Shortly thereafter, the organizations combined as the Air Force Cambridge Research Laboratory (AFCRL), becoming highly specialized in radar, digital communications for radar, geophysics, and weather analysis systems.

The AFCRL quickly established itself as a premier research facility, especially to William Golden, President Truman's special consultant on how the country should most effectively mobilize its scientific resources. After his team conducted a site survey, Golden reported back to the White House that "They regard the Air Force Cambridge Research Laboratory as probably the best of the military laboratories, and ascribed this to the management of Mr. Marchetti."[80] Northrop Aviation on the West Coast undertook celestial navigation research for its Snark long-range missile, as did Willow Run using analog computers, while at MIT under Charles "Doc" Draper, inertial navigation was being hotly pursued as a possibility. Another idea derived from German technology, wrote Stever, inertial navigation had not yet arrived in 1946, but it would eventually win out by 1953 or 1954.[81] "Remember that the seed comes first; if you are to reap a harvest of aeronautical development," Arnold had told his audience in 1937 at the Western Aviation Planning Conference. Arnold's *seeds* were abundantly sown and later reaped huge harvests in radar, computing, navigation, and the gigantic Air Force missile programs like Atlas, Minuteman, and still later both the Mercury and Gemini launches.

Ivan Getting, as president of the Aerospace Corporation beginning in 1960, witnessed Arnold's legacy take flight. "It was the Air Force, by modifying the Atlas missile," said Getting, "that first put up meaningful satellites . . . it was the Air Force Titan that followed with bigger payloads . . . [and] NASA didn't launch the Mercury or Gemini astronauts . . . they were launched by the Air Force."[82]

It appeared quite evident throughout 1946 that there was a lot of jet-propelled metal going skyward, and it didn't take a crystal ball to see that the future would bring forth more and not fewer jet aircraft, jet bombers, missiles, and rockets. New, very fast and very deadly weapons were preparing to enter America's military arsenal. On the other side of the world, the Soviet military was similarly occupied developing a comparable arsenal. In the event that one or the other of the adversaries started hurling metal, it would have been a real benefit to locate, track, and maybe destroy all or some of these jet-propelled objects well before they did any damage.

The 1946 battle plan for the United States, called Operation Pincher, didn't offer much of a solution to that problem; in fact, it was fairly threadbare of details. The basic plan was to abandon Europe and to retreat before the massive Soviet Army. Temporarily, the US would also abandon the Mediterranean Sea. US B-36 bombers flying from England and Turkey would then counterattack 20 select Soviet targets with atomic bombs.[83] The US Navy would then re-enter the Mediterranean and blockade the USSR. That plan preceded by three years America's acquaintance with the handiwork of Igor Kurchatov, the father of the Soviet atomic bomb. After the Soviets detonated their first atomic bomb, plans like Operation Pincher were goners.

In the meantime, the real sitting ducks in any US battle plan were the very vulnerable US Navy surface ships, especially aircraft carriers. They were vulnerable not so much from other surface ships or even submarines, but from above. Missiles and rockets fired from enemy aircraft prowling at some distance, standing off from their quarry and then letting go, would be particularly nasty. The Germans had already pulled off such attacks as early as 1943 and their technology was well known to post-war armorers on both sides.

German *lenkbomben* (guided bombs), either jet assisted or as winged gliders, were dropped miles from an intended target and would hit a ship as if coming out of nowhere. Between August 1943 and June 1944, a dozen destroyers, cruisers, and merchant ships were either sunk or badly damaged by what today would be called "robotic, stand-off bombs." The Henschel 293 was a bomb with a rocket strapped beneath it, like an early form of a cruise missile. It was unstoppable and deadly unless shipboard radar could spot the weapons in flight and attempt to jam any that were guided by radio. By 1944, the Germans had jam-proofed them. The most unfortunate World War II victim of a robotic bomb was the British troopship *Rohna*, sunk with 1,138 American troops on board. More unfortunate would have been the Henschel 293 in quantity flying with the Imperial Japanese Air Force. It never happened, but it would have given a whole new meaning to *kamikaze* attack.

The United States had its own versions called "glide bombs," some fitted with TV cameras for guidance like the GB-4, difficult to use but accurate to within 60 meters of its target; another, the US 1,000 lb radio-controlled, VB-1 Azon, was accurate to within 20 meters. With such "stand off" weapons, the Soviet Air Force, commanding upwards of 10,000 aircraft, might well have made short work of any navy attempting a blockade. In 1946, the US Navy was well aware that a small piece of America, about the size of an aircraft carrier, out in the middle of the ocean on a dark night was not a very safe place. The Navy had a good idea defending with surface-to-air Bumblebee and air-to-air Meteor missiles, but what good are they when 100 aircraft carrying several standoff glide bombs approach the aircraft carrier? How does the Navy keep track of 200 glide bombs at once? A handy gadget for the Navy to have in such situations might well be a control computer, together with a couple of bright boys to show them how to use it.

The Navy already knew about using a computer for both defensive and offensive operations, because it had asked Forrester about "applications of high-speed, electronic computation," to which Forrester replied with lengthy specifics on the military uses of computers in a letter in late January of 1946. The letter to Lt. Commander K. C. Knutson of the Special Devices Division elaborated on what Forrester called a "coordinated Combat Information Center . . . for rocket and guided missile warfare . . . automatic radar tracking and fire control . . . aerial and submarine torpedoes . . . [and the] stability and trajectories of guided missiles."[84] Unfortunately, the Special Devices Division was diminished into the Special Devices Center and eventually disbanded. Its control over computers was transferred to the newly formed ONR and its Mathematics Branch. The fate of Forrester's vision for a Combat Information Center, like Mendel's breakthrough paper on genetics that was first ignored by science and then lost for 40 years, slipped unrecognized beneath the surface and disappeared.

Chapter Five

1949

It is certain that Europe would have been communized and London under bombardment some time ago but for the deterrent of the atomic bomb in the hands of the United States.

—Winston Churchill, Boston Garden 1949

On a scorching September afternoon at the Pentagon taxi stand in 1949, George Valley dashed to the running board of a moving taxi. The lone passenger inside, a good friend, the Caltech theoretical physicist H. P. "Bob" Robertson, had hailed him. As Valley settled down inside, Robertson wasted little time getting to the point: "George, I've lately been briefed about the state of the radar system that the Air Force thinks it's setting up for air defense, and—it's scandalous! It's disgraceful! Why don't you do something about it?"

"Me?" replied the stunned Valley.[1]

As Robertson explained it, the country needed a way to defend itself from air attacks. Such a defense would undoubtedly need to be radar based, together with some sort of companion electronic system made up of who knows what coming from who knows where. He didn't have much in the way of specifics except that it must be a continent-wide, early warning radar system capable of spotting, tracking, and intercepting enemy bombers . . . and, oh by the way, all that defending and intercepting had to happen in real time otherwise it was useless. And the whole shebang needed to be built like yesterday.

The *continent wide* part of the request was as astounding as were his urgently vague thoughts on the requisite technology to pull it off. By continent, Robertson was talking about a nontrivial 247,000 miles of coastline enclosing some 14 million square miles of land that 152 million people called home. The United States had nothing remotely capable of taking on the task of such a massive defensive umbrella. But the need for one had suddenly been made clearly and terribly evident. The previous month, on August 29, the Soviet Union detonated a 20-kiloton atomic bomb in Kazakhstan. On September 3rd, a US weather plane picked up evidence of radioactivity off the coast of Siberia. On September 23rd, President Truman would announce to the public the explosion of the first Soviet bomb, four years earlier than expected. US intelligence would then uncomfortably add that the Soviet Union's intercontinental TU-4 bomber was quite capable of carrying its awful payload to the heart of the American homeland. In a scant 30 days America's once friendly skies would become suddenly full of dread.

Valley knew Bob Robertson well enough to know that he was no hysteric and that his dramatic outburst was not scary hyperbole. During the war they had worked together and had come to know one another well on OSRD assignments in London. Robertson was head of the scientific intelligence team that focused on the German V-2 rocket problem. He and Guy Stever were famous for their techno expedition off the beachheads at Normandy, following directly behind advancing troops to locate what rocket technology they could at V-1 and V-2 sites in France and Belgium.[2] Radically changed by his wartime experiences, Robertson stayed on in Europe well into 1946 working as chief scientific advisor for the Supreme Allied Command. Later, returning to teaching at Princeton, he continued government work as technology advisor to the Joint Chiefs of Staff, was instrumental in setting up the Weapons Systems Evaluation Group for the Secretary of Defense, and then served as its first director. If there was anyone well grounded about the extent of the Soviet threat, it was Bob.

Although only passingly acquainted with air defense matters, Valley had the unsettling feeling that the alarm in Robertson's voice was all too real. What he did not know while sitting in the taxi was that his friend's request would eventually come to occupy nearly every minute of his life for the next seven years. From this unexpected encounter and conversation between old friends, there would ensue an intensive and often frenetic manhunt for the people, technology, and plans to defend the skies over the continent of North America. It was a search the likes of which the United States had never experienced. Somehow the first thing to be done was to secure the necessary technology, for which neither of the friends knew where or how to begin to look. And once that technology was in hand, it would then need to be fashioned into the largest and most complex engineering project in

H. P. "Bob" Robertson (far left), with other Princeton faculty: Eugene Wigner, Hermann Weyl, Kurt Gödel, I. I. Rabi, Albert Einstein, Rudolf Ladenburg, J. Robert Oppenheimer, and Gerald Clemence, 1949. (Photo courtesy of the estate of Howard Schrader.)

postwar America. The enormity of the eventual project, in terms of magnitude, physical size, and money spent, is nearly always equated with the World War II atomic bomb-building behemoth, the Manhattan Project. In many respects it was far more complex and far more important. The Manhattan Project helped to end a war; the air defense of North America might prevent the end of a world.

George Valley, calm and easy-going, was a well-known, Washington-savvy, MIT physics professor and a member of the Electronics Panel of the US Air Force Scientific Advisory Board (SAB). A tireless worker, articulate with an incisive intellect, the well-degreed 36-year-old already had a half dozen years of military R&D under his belt. He had worked at MIT's Radiation Laboratory throughout the war and is credited with the development of the H2X radar bombsight for all-weather bombing, which played a strategic role in Air Force success. Returning to MIT in 1946, he edited four of the *Radiation Lab Technical Series* (collections that preserved for future researchers the important wartime technology developed at the Rad Lab). He also taught physics, and together with colleague Bruno Rossi continued his work on cosmic rays. As Valley recalled later, "No one expected a hostile country to possess nuclear bombs for years to come." America felt comfortably safe nestled between its vast oceans. "Cocksure and arrogant, most of us were fooled."[3] Valley quickly arranged to visit a radar station installed by the Continental Air Command (CONAC), remarking that, "The site resembled one of those army camps of the Indian wars that you see in late-night movies—except Quonset huts substituted for the log cabins and jeeps took the place of horses."[4] And the SAB supplied him with a stack of reports on air defense, which he called, "all disquieting." George Valley decided to act. It seemed Robertson had chosen well.

Equally disquieting as the bomb itself was the realization that the world's newest atomic gunslinger was Joseph Stalin. (Stalin was not his real name but a taken name that translates to "man of steel.") Treacherous and demented, Stalin was a one-man genocide machine, who, according to Robert Conquest's statistics in *The Great Terror*, was directly responsible for the murders of 8 million Russians.[5] The prevailing wisdom reckoned that if Stalin had such little regard for the lives of his own people, imagine what he was capable of on foreign soil with an A-bomb in hand. Equally crafty and bold, this despotic "man of steel" was the Soviet Union's absolute ruler beginning in 1929, and was on an expansionist roll since the Potsdam Conference in August of 1945. He pounced on the Baltic, absorbing Latvia, Lithuania, and Estonia, as well as the old East Prussian enclave of Konigsberg, renamed Kaliningrad. With Poland, Romania, Hungary, and Yugoslavia already under Soviet hegemony, he pulled off a Communist coup in Czechoslovakia in 1948. At Potsdam, he militated to share in the occupations of former Axis allies Italy and Austria but was rebuffed by the Allies. Failing that, he solidified the Kremlin's grasp on East Germany, and in 1948 blockaded Berlin; it took the Allies (mostly American) 270,000 cargo flights of supplies from 1948 to 1949 to break the blockade. Keeping Stalin from encroaching further into Europe was undoubtedly the steely glare from Harry Truman, who until August of 1949 was the world's only atomic gunslinger. Truman drew down twice with Hiroshima and Nagasaki, and seemed inclined to do so again if Moscow messed with him.

Stalin looked East to keep up the pressure. The Red Army had chased the Japanese out of Manchuria (an area the size of Western Europe), and then all the way down the Korean peninsula and out to sea, effectively liberating Korea after 38 years of Japanese occupation. More than a few Koreans cheered in delirious gratitude and tossed loving bouquets at their friendly Soviet liberators. By 1949, Stalin's Chinese compatriot Mao Tse-tung watched Chiang Kai-shek's battered army beat a hasty retreat to

Taiwan, which set the stage for Stalin's generals to put forth a battle plan for the North Koreans to race across the 38th Parallel in June of 1950 to reunify their recently partitioned country. Americans would know it as the Korean War. Everything seemed to be going Stalin's way. And the atomic plume rising over Kazakhstan gave him the leverage to strut across the world stage chortling about doing to the West what Mao had just done in the East. It's what put the hurry-up in Robertson's voice. Everyone was truly frightened; and the times were reminiscent of those unsettling months just prior to Chamberlain's concessions to Hitler. Did the offing hold another Munich-like appeasement, this time with the Soviet strongman, followed by another sneak attack on the United States? The Pentagon thought so.

On March 31, 1949 Winston Churchill addressed a full house at the Boston Garden.[6] He spoke of war's end and the prospect of peace in a prosperous future; he spoke of technology's gifts of radar and air power girding England during the Blitz. He quoted Tennyson's hope for people "earnest of things that they shall do;" and asked those before him, "Is time on our side?" to thwart the new menace rising up around us. Although the world was far from settled and peaceable, Churchill and his audience that day took comfort in the knowledge that America's nuclear edge kept the monster away. Like George Valley, they too felt sure that the Soviets were years from attaining atomic capability. None had any inkling that Kazakhstan was a mere five months off. And time was certainly not on their side. Klaus Fuchs, the German scientist and Soviet spy working on the Manhattan Project, had already funneled to the Soviets "the principle design of the plutonium bomb."[7] With that information in hand, Igor Kurchatov, the Soviet A-bomb master builder, and his understaffed researchers, bypassed years of blind alleys, preliminary experimentation, hundreds of millions of dollars, thousands of man-hours of labor, and tons of raw materials that a war-ravaged Soviet economy and industrial base would have found near impossible to equal.[8] They called it "development by imitation." All the Soviets had to do was follow the American bomb recipe, test its parts, and then assemble Arzamas-16 for detonation.

Fuchs' espionage was long-lived and thorough. In his confession to the US War Department and the FBI in January of 1950, he admitted that beginning as early as 1942 he had easy access to all relevant files and used them to provide the Soviets exacting specifications and drawings about the size of the bomb, what it contained, its construction, and how it was detonated.[9] Philip Morrison, an American physicist and a leader on the Manhattan Project, remarked upon seeing copies of the espionage, "this is the real thing."[10] Of course, Russian nuclear physics was highly advanced as far back as the 1930s, and the Russians would have ultimately, after a considerable technical and financial struggle lasting some years, achieved an atomic bomb on their own. However, imitation via espionage—easy espionage at that—was just too good a deal to pass up. The Russians even copied the bomb's exterior shape to ensure a neat fit into the bomb bay of their copied B-29, the TU-4. Nikolai Alexandroff summed it all up in the journal *Soyuz*, "There was no *Russian* atomic bomb. There was only an American one, masterfully discovered by Soviet spies."[11] At the time of their summer taxi ride, Valley and Robertson had no idea of how the Soviets had pulled it all off. It was beside the point. Their grave concern was how now to defend against it.

The Anvil of Complexity*

If a sense of urgency pervaded the backseat of Valley and Robertson's taxi, a decided pall of impending doom hung about the interior of the Barta Building. Not far from Churchill's podium, two miles or so across the Charles River and up Massachusetts Avenue into Cambridge, the answer to the air defense dilemma sat under the fretful gaze of Forrester and Everett. In 1948 they had moved from Gordon Brown's Servomechanisms Laboratory into the Barta Building on Massachusetts Avenue, a block from the front door of MIT. During the move, they jettisoned their work on the flight analyzer (officially, the Aircraft Stability and Control Analyzer). It was a major part of their contract with the Office of Naval Research, and their decision did not sit well at all with the ONR. It had not seemed too rash a decision at the time. After all, building an electronic digital computer, as they had witnessed at the Moore School, was no trivial task. Pursuing a single machine, the more important and practical of the two, made real sense. Years later, wouldn't Herman Goldstine tab ENIAC as the most complex machine ever built by man up to that time? Well, Whirlwind was similarly complex and as such, they felt it needed their fulltime attention.

The problem in 1947 was that hardly anyone realized how difficult it was to build an electronic digital computer. Forrester and Everett were convinced that simultaneously trying to build both computer and simulator might have jeopardized both. The decision was full bore ahead with the behemoth computer only, which they reckoned was plenty enough of a challenge. However, eliminating the flight analyzer had its consequences, because it effectively eliminated their ONR mission as well as a Navy contract

---

*See Chapter Five notes, Section A, for information on Kent Redmond and Thomas Smith, chroniclers of Whirlwind and SAGE, whose 25 years of related research is used extensively.

stipulation about building a computer-controlled flight analyzer. Their decision, although it may have displayed rational thinking and sensible engineering, displeased some and was used against them by others.

Tension over the flight analyzer was not their only problem. Barely six months after returning to Cambridge, they began to fend off repeated attempts to impede some or all of their progress in the planning and building of Whirlwind. The years 1947 to 1949 were times of ever-increasing trial and turmoil for them and their youthful mates. Nearly all of the discord concerned either money—that they were incorrigible spendthrifts—or assailed their credentials—that they were a bunch of no-name, junior engineers without any standing. Somehow Forrester and Everett had latched onto a big fish that they did not really deserve. People more senior than Forrester and Everett viewed themselves to be far more deserving of the government's largesse. Flamm ascribes the flurry of academic rancor largely to money and bruised egos, what he calls "the galling to established members of academic mathematical circles to be in close competition for research dollars with a group of young, largely unknown MIT engineers, with no real finished product to show for their efforts."[12] Of course, such pettiness was all true. But if youth, lack of reputation, and zero previous success were removed from the equation of innovation and discovery, there would be pity few inventions in the world today.

Simultaneously trying to defend themselves and to prove themselves took diplomacy and guerilla instincts. Forrester and Everett were ever watchful as to who or what was to come at them next with either criticism of their project or a serious suggestion to shut it down and scrap the hulking giant of a machine. Hardly a month went by when someone or other from the Office of Naval Research (ONR) in Washington, the ONR's local Boston office, or someone else at the request of the ONR's Boston or Washington offices, or from the ONR's Special Devices Division, or from MIT itself, did not sweep into the old laundry building to call into question the budget, the plans, the personnel, the materials used, the project's mission, its ultimate

Exterior of Barta Building, 211 Massachusetts Avenue, cicra 1949.
(Photo courtesy of MITRE Corporation.)

utility, or its completion date. Adversarial reports were filed against them, to which they had to reply in writing and at length, sapping away valuable time and emotional energy better spent with staff or on their beleaguered computer. Secret letters circulated that tainted them, their efforts, and their machine; letters to which they were not privy, and as such never had a chance to debunk. A decided buzz of suspicion hovered over their every move. Odd happenings were taking place, and odd behavior coming from people who should have known better and probably did. By late in 1949, the craziness was all coming to a head, and an axe was somewhere getting a sharp edge to separate the bright boys from their machine.

A tragic end seemed looming for something that had begun so grandly. From the project's outset, Forrester's innovation and deliberate organizational moves seemed shrewd, professional, and above any reproach. He set up his laboratory for ten divisions: seven for technical research and three for support. With the help of Nat Sage, the well-respected and highly experienced director of MIT's Division of Industrial Research, he assembled an extraordinary staff. "The general type of man whom we need," he explained to Sage, who was to cast about for warm bodies to fill each slot, "should have originality and what is often referred to as genius. He should not be bound by the traditional approach."[13] Everett headed the all-important Block Diagrams Group, which would "ascertain machine computing techniques, programming techniques, and component designs for accomplishing computing, storing, switching, and programming."[14] Two more bright boys, Stephen Dodd and Pat Youtz, would work part-time with Everett. Other bright boys enlisted to the cause included Harris Fahnestock, in charge of Administration, and David Brown, Electronic Engineering.

Putting the simplest things down on paper helps to visualize them, putting a highly complex, never-before-attempted machine on paper was a must. Everett's Block Diagrams Group worked its way through the tricky aspects of knitting all the essential components of Whirlwind together before committing anything to assembly. Some things were simple enough, like figuring out the need for standard electronic and relay racks accommodating removable bases 17 inches by 10.[15] Sylvania was nearby and could easily fill the need. Other things were more subtle, and the Diagrams Group let out a collective sigh of relief each time a dangerous one was encountered and not yet built. Such was the case when first diagramming Whirlwind as a serial computer (serial computations are done one at a time in succession) as opposed to a parallel computer where more than one computation can take place simultaneously. "Block diagrams that Everett had developed for serial transmission—inspired originally by the EDVAC machine—had convinced the engineers that, despite relatively simple and easy-to-maintain circuits, such a device would be slow."[16] Too slow for the real-time computer that they intended Whirlwind to be. Much more diagramming and rediagramming took place, and many more sighs of relief issued up from the bright boys as they plodded along from tube to connector to circuit to voltage. "By 1947, the logical design of Whirlwind had been completed."[17] On September 4, 1947, the 132-page Report R-127, *Whirlwind I Computer Block Diagrams*, was submitted to the Navy's Special Devices Center. The bright boys were succeeding, and grandly so. They were young, brilliant, confident, and cocky, which for many hovering nearby waiting impatiently for Whirlwind to go bust, was grandly irksome. There were remarks that the bright boys were "arrogantly high-hat and snobbish, working in a building closed by security regulations to outsiders, that they were as unrealistic about what they were doing as they were young and immature, and theirs was a gold-plated boondoggle, extravagant in its demands, in its rewards, and in its raids upon the taxpayers' purse."[18]

Lost completely on the ungracious onlookers was the fact that the bright boys were fast about constructing one of the most complex machines ever created by human beings and certainly the most complex ever created at MIT. The bright boys were at the anvil of complexity hammering out an enormously intricate machine complex piece by complex piece and hoping that when the many thousands of parts were finally assembled and the switch thrown that Whirlwind would crank up and be ready and able. Forrester and Everett knew how careful Presper Eckert had been with every aspect of ENIAC. Eckert had even tested electrical wire coverings on a cage full of Moore School rats to see which they favored most and least. Little things can mean everything when working on an electronic computer. If anything goes wrong, even a seemingly small mistake can be disastrous, and may be all but impossible to find and correct. Forrester argued that taking precautions took a "large part of our time and cost."[19] He noted that unlike a radio that "could accommodate some static, a television some flicker, and a teletype some error without destroying a message . . . a digital computer could tolerate no such error without destroying the message . . . a single error could produce chaos."[20] He and Everett insisted that all parts, assemblies, and subassemblies be tested individually and then together, and that designers should take the time to build whatever test equipment was necessary to locate deficiencies and remedy them before building them into Whirlwind.[21] And build nearly everything they must; they couldn't nip out to the nearest electronic supply house for their computer gear, there

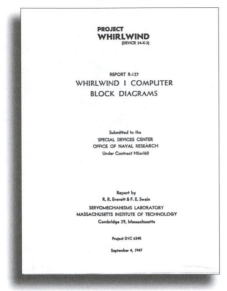

PROJECT
WHIRLWIND
(DEVICE 24-X-3)

REPORT R-127
WHIRLWIND I COMPUTER
BLOCK DIAGRAMS

Submitted to the
SPECIAL DEVICES CENTER
OFFICE OF NAVAL RESEARCH
Under Contract N5ori60

Report by
R. R. Everett & F. E. Swain
SERVOMECHANISMS LABORATORY
MASSACHUSETTS INSTITUTE OF TECHNOLOGY
Cambridge 39, Massachusetts

Project DIC 6345
September 4, 1947

Front cover of Report R-127, 1947.
(Image courtesy of MIT Archives.)

weren't any such supply shops or any such gear. They had to rummage for parts or have them specially designed and built.

Everything about Whirlwind was new, untried, and difficult. And very quickly the bright boys found themselves dealing with a previously unheard of level of complexity—pioneering it. It was like a newfound land to colonize and civilize. One false step and they would be lost, subject to the hee-haws of every doubting jackass from Cambridge to the Whistle Factory. Only their brethren from the Moore School lectures, off on their own somewhere hacking through similar dense stands of the unknown, could appreciate their plight. "We were engaged in an endeavor," said Forrester, "that no one else could claim to understand."[22] Gordon Brown, their boss in the Servomechanisms Laboratory, realized it, and his benign neglect gave his bright boys time and space for their thoughts to rattle about and gain enough composure to come through the experience. "We saw Gordon infrequently, maybe once a year," Everett later recalled, warmly thankful for Brown's hands-off approach. "Otherwise he left us to our own devices."[23]

Complexity not only engaged them with their machine and its intricate relationship of parts and functions, but also dwelled within each of them. Immersed in complexity as they were, they began to think in complex patterns and share complex notions. They peppered their engineering journals with the process: self-doubt, false starts, and then hurrahs of delight at every success, even the smallest, before finally the new knowledge settled in for good and became a part of them. "There are vast realms of consciousness still undreamed of," said D. H. Lawrence in "Terra Incognita," his homage to the unknown, "vast ranges of experience, like the humming of unseen harps, we know nothing of, within us."[24]

Constantly thinking about complexity, the myriad parts of their machine, its systems and subsystems, the changing relationships among the parts as an electrical force was applied to them, and the resulting actions and reactions taking place within and among the system because of that force, opened doors of insight and perception that were as new and powerful to them as their new machine. And as Lawrence concludes, like the "slow mounting marvel of a little tree" complexity began to reveal itself, bit by bit taking up residence in the intellect with its newfound enlightenment. The thesis trail shows how important having the machine was to the journey. Perry Crawford, the digital proselytizer who had hooked Forrester in 1945, wrote in his master's thesis, "Automatic Control by Arithmetical Operations" that "It is to be emphasized at the outset that little progress has been made toward the construction of automatic electronic calculating systems for any purpose."[25] Written in 1942 and predating ENIAC, Crawford's thesis stood with its toes nudging at the borders of the unknown land, but went no further.

Like Dr. Who, the intrepid time traveler from the British TV classic by the same name, who jaunts through time in his TARDIS (Time And Relative Dimensions In Space) vehicle, Crawford also needed a special vehicle to gain entrance into his new world. He needed an electronic computing machine. He hadn't one because none existed, so his thesis is filled only with wonderful speculation. Bright boy, David Israel, was luckier. He had his TARDIS in Whirlwind and with it leapt far forward in, "The Application of a High-Speed Digital Computer to the Present-Day Air Traffic Control System" (1951). "The use of these new machines," he wrote, "as the central or directing elements in complex physical situations is similarly possible and desirable, and in such applications those machines offer great promise in permitting humans to more fully exploit and utilize recent advances in technology and science."[26] Israel opened the door to future skies where aircraft would be closely watched over and monitored, with every flight path controlled by a computer.

The bright boys and Whirlwind announced a new integration of man and machine. And very much like Dr. Who's TARDIS—that strangely was larger on the inside than on the outside—Whirlwind too was similarly larger on the inside than the outside. There were worlds within it. Israel's brilliant thesis beckoned others to follow to see what else these new machines would serve up and where they might take us. Up the road a ways from the bright boys awaited Licklider and "Man-Computer Symbiosis."

Things that don't kill you make you stronger, they say, and the bright boys packed on some muscle from their journey into complexity. They would need it to heft about Whirlwind's chunky blueprints, Report R-127, *Whirlwind I Computer Block Diagrams*, which Yaneer Bar-Yam asserts in *Dynamics of Complex Systems* is the "quantitative measure of how complex a system is . . . [that its] complexity depends on the level of detail required in the description."[27] The vast, intricate system that was Whirlwind was very well endowed with details about itself: mainly, a year's worth of them in the stout volume that was R-127. The bright boys might not have had standing with senior academicians or have authored scores of papers and textbooks, but they now possessed a deep knowledge and skill the likes of which no one could challenge.

Better yet, they knew how to put that knowledge to work. It might not be enough to save Whirlwind, but they possessed much to be reckoned with. They had a tiger by the tail and were riding it for all it was worth. "We were cocky," reminisced a bright boy years later. "Oh, we were cocky! We were going to show everybody. And we did."[28] And only the best would do for them as they transformed their patch of complexity from *incognita* to *cognita*. They felt entitled to their seeming budgetary extravagance, saying that it "was a deliberate policy of saving development time and money in the long run by insisting on going first class."[29] Many of the bright boys had witnessed the very same extravagance working its magic at the Rad Lab during the war. The Rad Lab environment was an incubator of wonderful discovery . . . but for a price. Couldn't the bright boys ape that success; replicate in a small way the Rad Lab experience as closely as they could?

Be careful what you wish for, because the Rad Lab was not immune to outside pressure. There were forces, even during those fearful and uncertain times of worldwide warfare, which tried to disrupt and to hijack research. The historian Henry Guerlac interviewed Vannevar Bush on that very subject during the war (August of 1944) and got an earful on some of the Rad Lab's "rip-snorting" battles with big companies wanting a piece of the action. "Dr. Bush spoke of the Big Companies as being 'damn conceited' but the RL [Rad Lab] physicists were 'also conceited' . . . and there were times when he felt like saying 'a plague on both your houses' . . . but, he concluded, 'the stuff never would have been done if the Big Companies had done it.'"[30] It seemed that if properly insulated from without and with a budget to get the job done right the first time, a dedicated laboratory staffed with the young and the brilliant trying hard to prove themselves had a fighting chance at pulling off a big win.

Forrester remembered well that wartime budgets were near limitless and made spendthrifts of everyone, but they had worked wonders. Could he and the boys pull off their own small miracle? He felt sure that $100,000 a month for Whirlwind might do the trick.[31] Savvy, strong-minded management also goes a long way at maintaining stability. The Rad Lab had Lee Dubridge as an excellent front man together with the keen eye and dogged energy of multimillionaire Alfred Loomis. The bright boys had Gordon Brown as mentor and guardian, who would eventually become dean of

MIT's School of Engineering and would soon after revamp MIT's engineering curricula from its "classic 'bluebook' series written by MIT faculty back in the 1930s."[32]

The day-to-day leadership fell to Forrester and Everett, both young (in 1947, 29 years old and 26 respectively) and untested on large management projects. Forrester would be the battle shield and mouthpiece for the group. His style worked, but it drew harsh criticism and from some, even personal animosity.[33] "Dedicated to Project Whirlwind and determined to secure its success, Forrester aggressively and single-mindedly pursued the course which he believed would most quickly reach that end."[34] Rather than arrogant and high-hat, he was strong willed and convinced "that the social structure and management of an enterprise are far more important than the underlying science. If one has the right environment, it will produce the required science. But the best science will fail in an unfavorable social and managerial setting."[35]

Sometimes too, the bright boys inadvertently or ingenuously shot themselves in the foot. Dropping the flight control analyzer part of the Whirlwind project in June of 1948 became the most obvious and easy miscue to use against them. But it was a flawed and hokey excuse at best, especially by the ONR. The Navy was well aware as early as January of 1946 that the bright boys saw Whirlwind as a general-purpose computer and not yoked to the bidding of the control analyzer. Forrester's letter of reply to Lt. Commander Knutson in 1946, responding to Knutson's request about Whirlwind's capabilities, made it patently obvious, months before the Moore School lectures, in what direction the bright boys' minds were turning. "The Navy Special Devices engineers enthusiastically realized that they were contemplating a revolutionary device . . . "[36] In 1947, while still diagramming their way through Report R-127, Forrester and Everett scurried up a bit of their spare time to churn out some extended reading matter for the Navy on the military uses of computers and radar. Known as Servomechanisms Laboratory Reports L-1 and L-2,[37] which to the Navy must have read like Jules Verne takes on the 6th Fleet, they contemplated how Whirlwind would assist in "a problem involving 10 ships, 5 submarines, interconnecting radar and sonar data, and depth charges in any number up to 20 pre-set units and 20 proximity-fuse units in the water at one time."[38]

Report L-1 concerned itself mainly with Whirlwind helping a destroyer to acquire target data for depth charges; Report L-2 dealt with combat command and control information and communication among ships of an antisubmarine task group.[39] L-1 and L-2 not only broke new ground for computers in general but also discussed topics totally unheard of prior to the reports. The bright boys were carving out the likes of a hitherto never realized computing machine and at the same time the potential for its use well beyond that of number crunching. For Forrester and Everett, "to the best of their knowledge at the time and in after years, [they] knew of no other practical engineering work on how the logic of computers could be applied to interpret radar data."[40] For the military, this view of Whirlwind and computing—there in 1947!—was eye popping. Did someone say, flight control analyzer? Who really cared anymore now that this new look at the future of combat had arisen?

The passage of time and technology rendered moot the need for the 1944-era flight control analyzer. The original de Florez concept of using a flight trainer and calculating machine for determining the handling characteristics of an airplane prior to its construction[41] was, just three years on, still a good idea but seriously dated. It was not the technology that was dated, because the technology needed in 1944 to build de Florez' flight analyzer still was not available in 1947. It was the engineering requirements for postwar flight analyzers that were very dated; prewar subsonic aircraft analysis was now supplanted by the transonic and supersonic. And all of it went way beyond the bright boys' capabilities; capabilities needed for displaying and solving some 90 equations in real time, like aerodynamic coefficients, instrument equations, equations for motion, and earth axes.[42] The necessary technology just wasn't there and wouldn't be for some time. The dramatic shift away from the subsonic was quite evident just across campus where MIT's Department of Aeronautics was fast about building competency in transonic and supersonic flight, including, with more Navy money, a new wind tunnel for testing aircraft in the Mach 2 and Mach 4 ranges.[43]

Whatever the bright boys had for a flight analyzer was by 1947 best used for something else; any future for any flight control analyzer would first need serious attention from specialists like aeronautical physicists and flight dynamics engineers. Robert Seamans, much later an executive director at NASA and even later Secretary of the Air Force, directed 120 staff on the Department of Aeronautics' flight dynamics project,[44] which was all about the future of jet aircraft and missiles. By contrast, the bright boys numbered at most 15 full-time research engineers and 50-odd graduate students. The bright boys were best served by keeping to their strengths, which they wisely did.

Whirlwind in 1947 existed nowhere except on paper as Report R-127, *Whirlwind I Computer Block Diagrams*. Its design called for it to be a real-time, general-purpose computer, with its real-time capability putting it into a class of one worldwide.[45] However, real time came with a cost, which Forrester and Everett understood and accepted.[46] The tradeoff was high-end accuracy, and because of the limitation Whirlwind could not meet the arithmetic functions needed for high performance analysis of modern airframes. Transonic and supersonic flight analysis needed a "special-purpose" computer able to crunch really big numbers really fast.[47] Unfortunately, getting the required real-time flight analysis from a special-purpose computer would take technology that just was not available. If Forrester and Everett had stuck with both the digital computer and the flight analyzer, they would have most likely overspent and underproduced for a very long period of time and in the end would have very probably been overtaken by the flight analyzer competition, which included the all-time heavy hitter, Edwin Link, the father of flight training and analysis. Not a pretty picture. The bright boys were invested in real-time, general-purpose computing and applying their machine to the visions they set out in L-1 and L-2. They were not about to jeopardize all that potential to limit their machine to a singular function like that of flight analysis.

The Navy would eventually get its special-purpose trainer/analyzer, research for which began in 1950 at the University of Pennsylvania. Ten years later in 1960, the fruit of that research, called the Universal Digital Operational Flight Trainer or UDOFT, rolled out of a Sylvania factory.[48] Soon after the UDOFT appeared, Edwin Link came out with his Link Mark 1. Forrester and Everett had dodged two bullets: the ten years that it took to produce the flight analyzer, and their narrow escape from locking competitive horns with Edwin Link. The problem then became *what next* now that the flight analyzer was kaput? The Special Devices Center no longer had a "device" to nurture into existence. Whirlwind, therefore, became, in the eyes of the ONR, just another computer. And like the ONR's other computer investments, it fell under the jurisdiction of the ONR's Mathematics Branch. Alas, the math guys were readying a last laugh.

It took a little gumption to nix the flight analyzer and stick to the real-time, general computer plan. It was a gamble that paid off. However, it took a lot of forbearance on the part of the Navy to accept major changes to a product on which they expected Forrester and Everett to make good. After all, the Navy had already fronted the bright boys nearly $1 million ($75,000 in December of 1944 and then $875,000 more in June of 1945) with no flight analyzer to show for the investment. Forrester then sold his Navy patron on going digital in 1946, which got him another fat contract and more time.[49] Forrester and Everett again switched up on the Navy by canning the flight analyzer altogether. The reasons for the switch were well known and agreed to by the Special Devices Center, but nonetheless, Whirlwind must have seemed like an embarrassing money pit.

The bright boys did, however, keep the Navy informed at every stage of their research. Forrester insisted on keeping voluminous laboratory status reports that were, as he called them, "as frank as they were frequent." At any time throughout Whirlwind's birthing, the Navy had complete access to every step of the process. The bright boys risked the Navy's money, everyone's time, and their shot at big-time success because they strove to build reliability into Whirlwind, which was the big knock against all of the early electronic digital computers. Reliability is what chewed up the Navy's investment the most. The few big machines that were in existence in 1947 were wonderful, but most pooped out after short runs. Many times they ground to a halt owing to rather minor malfunctions.[50] The bright boys aimed for Whirlwind to be a near flawless, highly reliable machine. And in 1947 their quest began to bear fruit. First out of the box was bright boy Norman Taylor who delivered on real-time computation with a small (6 feet high by 15 feet) testing machine tabbed the five-digit multiplier. Taylor got the machine to multiply two five-digit binary numbers in 5 microseconds with high reliability, and soon after got the machine to multiply for 45 days without an error.[51] That many errorless days of reliable operation was simply unheard of from any machine anywhere.

Another big trouble spot for all early computers was with their electron tubes,[*] most of which were used by the thousands in each computer as electronic on/off switches for logic circuits. They also had giant cousins, called electrostatic storage tubes, which were used for storing binary data, what we now refer to as RAM or Random Access Memory. In 1947 these tubes were considered state-of-the-art, predating the transistor, which was invented in 1948 but barely employed in computers until 1953. Electron tubes replaced the older and much slower technology of the Mercury Delay Line. A delay line was a five-foot tube of mercury (usually used in sets of tubes and referred to as lines) in which an electric pulse, representing a binary bit, was converted into a sound pulse, then traveled as a sound wave through the tube of mercury, emerging at the other end where it was reconverted into an electric pulse again. About a thousand binary bits could be stored in each mercury tube, but accessing that memory was slow—hence, the choice of an electron tube which was much faster. Today, of course, both electron tubes and mercury lines are long gone, replaced at first by

---

*See Chapter Five notes, Section B, for brief history on electron tube.

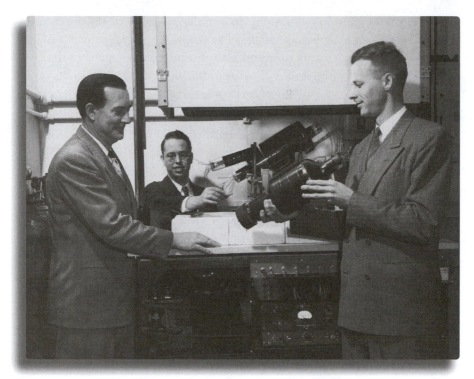

Whirlwind's on-site tube manufacturing department: (left to right) Pat Youtz, Stephen Dodd, and Jay Forrester examine finished electrostatic storage tube, 1951. (Photo courtesy of MITRE Corporation.)

integrated circuits or ICs populated by teeny transistors that acted as the on/off switches and RAM, and later still, the ICs were supplanted by microprocessors inhabited by millions of transistors, whose transistor populations seem to grow by the millions every year.

The problems with the electron tube switch and electrostatic memory were that they were expensive—especially memory tubes—short lived, and could blow out unexpectedly. Maintaining the reliability of electron tube switches, numbering in the thousands in a computer, was a frustratingly nasty affair. They aged quickly and badly; they leaked electrons, lost efficiency and then burned out. Burnouts caused computer errors that would shut a machine down until the tubes were replaced. "Radio engineers were not concerned that the life of a vacuum [electron] tube was about 500 hours," noted Forrester. "But computer engineers, considering the use of many thousands of vacuum tubes, easily estimated that with such a short life, the machine would run no more than a few minutes between failures."[52]

The electrostatic tubes, either commercially available or those custom built by bright boys Stephen Dodd and Pat Youtz, had extraordinarily brief life spans and "were costing us about a thousand dollars [each] to make," recalled Forrester. "They would store about a thousand binary digits and last about a month."[53] Resolving tube problems while fending off pressure from their frequent interrogators took considerable tact and calm. But they came through it. To counter imminent tube failure that would have rendered a computation utterly useless took a Forrester brainstorm to solve. Using Taylor's machine, Forrester invented marginal checking, which automatically spotted the tubes that were soon to fail. [54] Replacing those tubes early on gave them the fast switching and dependable memory that they sought for Whirlwind. Forrester realized, however, that marginal checking of electrostatic memory was merely a stopgap, that something more permanent would soon be necessary. He began a search for such a memory capability, and that quest would eventually lead him to one of the most important discoveries in computer history: magnetic-core memory.

"Chance favors the mind that is prepared," promised Louis Pasteur, and Whirlwind seemed to provide the mental preparation for all kinds of chances to spark into being. Sometimes, as with David Israel, it was how to use the machine to effect other machines or processes; sometimes too, as with Jay Forrester, it was a new advance in technology or the application of one technology to advance another. For bright boy Dudley Buck, Whirlwind's gas-filled, glass tubes and glowing filaments would launch the 23-year-old research assistant into the chilly world of superconductivity, which is the ability of some metals to conduct electric current with no resistance at extremely low temperatures, like those below minus 420 degrees Fahrenheit. Buck's eventual discovery of the cryotron, a superconductive computer switch, based on the effects of magnetic fields at liquid helium temperatures, represented the first practical use of superconductivity. A hundred cryotron switches could fit inside a thimble, a far cry from the space needed by a hundred electron tubes. Although Whirlwind never wore any of Buck's cryotrons, just the machine's presence on campus had an alluring power to grab a mind like Buck's and to switch it on.

In the years to come, Whirlwind would switch on many other minds. Robert Noyce for one, who in 1947 was still an undergraduate at Grinnell College in Iowa. He arrived at MIT in 1949 as a graduate student hell-bent with taking on electronic digital computers, and he would soon after co-invent the integrated circuit and then go on to cofound the Intel Corporation. A youthful Ken Olsen prowled the monstrous racks of Whirlwind and toted away the formula for a modular computer that became the hallmark of his equally monstrous computer company of the 1980s, Digital Equipment Corporation or DEC. There were indeed other worlds within the machine, *terrae incognitae*, "vast ranges of experience" yet to be discovered that were as important as the machine itself. And Forrester and Everett were its gate-keepers. Technology and fortunes beyond all imagining shimmered between the covers of Report R-127.

In 1947 the long-range hope for Whirlwind as a real-time, general-purpose computer rested with L-1 and L-2, and the hope that those reports would generate some attention and recognition in the right places. The bright boys' direction for Whirlwind was concerned with early warning, detection, interception, and command and control above and beneath the water, with sonar and radar data and communications all tied together. Such potential would make for joyful pandemonium in the halls of the Pentagon, if the news ever reached the right ears. Whirlwind was being crafted into the brain of an information system capable of keeping a close eye on the movements of, say, Stalin's hordes, or maybe of providing a bit of strategic forewarning about an impending attack. L-1 and L-2 might also divert for Whirlwind some of that rich flow of research money that seemed to be endlessly cascading into MIT from the military, especially into the Research Laboratory of Electronics. That early promise for L-1 and L-2 withered in 1947. The ONR missed a huge chance at shoring up a good chunk of national security. And it was exactly the kind of national protection that Congress had in mind when it created the Office of Naval Research in 1946: "to plan, foster, and encourage scientific research in recognition of its paramount importance . . . for the preservation of national security."[55] And when over time the Special Devices Division devolved into the Special Devices Center, and thereafter, as mathematicians from the ONR's Mathematics Branch took over Project Whirlwind, the cocky, mission-jeopardizing bright boys were in the hot seat for getting rid of the flight analyzer.

Blame it on Calculus

If you were any kind of a mathematician in 1947, you kept an eye on the whereabouts and doings of John von Neumann. He was a trendsetter. He hobnobbed with all the big people, events, and places in postwar America. He contributed to quantum physics, functional analysis, set theory, economics, computer science, numerical analysis, hydrodynamics, and statistics to name a few. He dressed well, drove automobiles poorly, and threw extravagant parties with his wife Klara at which he drank too much, ate too much, and joked until the wee hours of the morning. It was said of the great polymath that the only thing he could not count were calories

When von Neumann dropped Howard Aiken's Mark I electromechanical machine like a hot potato and dashed off to embrace ENIAC, and soon thereafter began building his own electronic digital computer, only fools missed the import of his newest of trends. Von Neumann recognized in a heartbeat that his calculations should and would only be run on an electronic digital computer. The reason was simple: they were the only machines that could do the job that he needed to have done. The trendsetter was an instant convert and passionate about the new world before him. When von Neumann first glanced at the bright boys' five-digit multiplier, it had just completed 5 billion multiplications without an error. He got so excited that he kissed Norman Taylor on both cheeks.[56]

Von Neumann's kiss was less in congratulations for the machine's ability to run without error and more a mathematician's exuberance at the prospect of freeing calculation from the drudgery of doing it by hand or, better yet, enabling solutions to previously intractable problems. A finger-numbing example of the drudgery is a famous calculation from 1947 by the illustrious mathematician and statistician George Dantzig (although wisely he had no "hand" in the actual calculation himself). It's what Licklider would later consider a woeful waste of time and resources. Dantzig was an expert at what was then known, especially by the military, as linear programming, which in 1947 meant planning methods solved by desk calculators. One of his first applications of linear programming was the determination of an adequate diet that was of least cost (something valuable to know when figuring out how to provision troops). It was a problem with nine equations in seventy-seven unknowns. "The solution required nearly 17,000 multiplications and divisions, which were carried out by five statistical clerks using desk computers [calculators] in 21 working days."[57] Twenty-one days! Today such an exercise simply boggles the imagination, except maybe there is some small comfort in the fact that five people were gainfully employed for four weeks because of all that keystroking. Magnify those five keystrokers by the hundreds and Richard Sprague's Northrop Aircraft plant rises into view. "Try to imagine," he recalled, "a room full of engineers, designers, and computation personnel, stretching as far as the eye could see, all operating adding machines, desk calculators, and comptometers. That was the aerospace computing center in 1945."[58] And he called it a "computing center."

Hollywood portrayed such grim toil in the movie *The Apartment* (1960) in which Jack Lemon played Bud Baxter, a comptometer clerk for a huge Manhattan insurance company. In the opening scene Baxter sits at a bare desk with a toaster-sized calculator in front of him. The huge chunk of metal bristling with rows of menacing keys is his workday purgatory. Surrounding him and stretching back for what looks like a hundred yards or so is a sea of other bare desks with similar calculators. Others just like him sit at those desks—an army of slavish workers who punch numbers into the metal monsters eight hours a day. The racket of clacking calculators is deafening. And what happened to the scores of calculations eventually produced amid all that thunder? How did one clanking toaster make its answer known to another toaster? By hand! Baxter and his mates hand carried sheets of calculations from one to the other or had them picked up by messengers and toted away. A von Neumann kiss was what the future would plant on these electronic, digital computers as the new machines made bleak rooms of comptometers disappear forever.

Bright boy Jack Harrington would put an end to such isolation of machines when he taught Whirlwind how to pass information around electronically. But first the new machines had to properly introduce themselves, to gain "closure," as *The Social Construction of Technology* calls it.[59] Electronic, digital computers had to somehow achieve that closure gained only through social acceptance, an acceptance necessary for the machines to finally flourish. The message that the gears had ground to a halt at the Whistle Factory now needed forceful and wide dissemination. The bright boys did their talking with their machine. Whirlwind was loud. It would have to be to out shout the entrenched analog computer experts who felt threatened by these upstart electronic computers.[60] The analog computer-servo crowd was deep with engineers who plied a brisk wartime trade of remote-controlled cannon and automatic pilots and who, quite naturally, figured the same technology would, postwar, glide smoothly into factories and become, as George Valley called them, "humming hives"[61] of analog.

Most people who were aware of the new, all-electronic digital computer looked upon it as a crass pretender to the analog crown and would persist in that notion for a handful of years to come; others dismissed it outright as a specialized tool soon to be marginalized to labor only on specialist tasks.[62] "Specialist" as in maybe a substitute for Dantzig's five statistical clerks and their hand calculators. The Senior Tom Watson, chairman of IBM, famously remarked, "I think there is a world market for maybe five computers," which was surely pigeonholing the machines to specialist careers only. IBM punch card calculator systems were so much in demand that Watson's factories were working three shifts to keep up with the orders. He said, "IBM is an institution that would live on forever, based on the punch card."[63] But when ENIAC appeared on the front page of the *New York Times*, the first such device ever to grab such an exalted slice of media attention, it made, as Herb Grosch remarked, Watson Senior "grind his teeth so ferociously that people, dozens of people, shivered as far away as Endicott, home of the IBM research laboratory." He immediately countered with the spare-no-expense, fast-or-else building of the largest and last of the electomechanical calculators, the giant Selective Sequence Electronic Calculator or SSEC.

Watson proudly displayed the SSEC in the heart of Manhattan, putting it in the front window of IBM World Headquarters on Madison Avenue where it sat from 1948 to 1951. Von Neumann graced the huge beast with a problem that had gone unsolved for 64 years because it would have taken 100 years to solve by human computers with their hand calculators. Approximately 150 hours later the SSEC produced the answer. The SSEC was way too much way too late, but it quite effectively warned IBM not to bet the farm on the punch card calculator business. Watson's son, Tom Jr., home from wartime service, digitally enlightened, and just easing his way back into the firm, would have his work cut out edging dad into the new world. *Popular Mechanics* magazine chimed in with a 1949 prediction that the machines would slim down from the elephantine: "Computers in the future may weigh no more than 1.5 tons." Seems that opinions good, bad, or indifferent were beginning to crop up about the machines, which is the surest sign that they were starting to grab some serious attention. And later, when *Time* magazine honored Howard Aiken's part electronic, part electromechanical hybrid, the Mark III, with its cover, popular fascination began to buzz. Aiken did not seem as confident in his offspring's future as had the forecast from *Time's* cover story, waffling that "there will never be enough problems, enough work, for more than one or two of these computers."[64]

At MIT, with the bright boys just across campus, the realization arrived early that this was no mere pretender. Painfully aware that a potential threat was arising and prepared to resist it was engineer Sam Caldwell, head of MIT's Center for Analysis. In the 1930s, with a $45,000 grant from the Carnegie Corporation, he and Vannevar Bush began the Center and built Bush's analog gear works, the Differential Analyzer, which would become world famous in a hurry and elevate MIT to the pinnacle of computing supremacy. Many of the bright boys, including David Israel, had studied under Caldwell; his signature even adorns Perry Crawford's master's thesis. Caldwell was quite cognizant of the ebb and flow of computing technology.

Many of his students trotted on over to classes in digital computing taught by Gordon Welchman, formerly of Hut Six and Enigma code-breaking fame with his partner Alan Turing. Caldwell was keenly aware of the gathering momentum of electronic digital computation. However, seeing was far from believing. He persisted in doggedly shoring up the sand castle as the tide rushed in and around him.

In a letter to his staff in early 1947—as Whirlwind's Report R-127 was unfolding—Caldwell laments the passing of the power and prestige of his Center but is dismissive of the new machine on campus, writing that " . . . in the field of electronic computation we entered the war among the leaders and emerged in a much less favorable position . . . resumption of our work in electronic computation development, at a greatly increased rate, stands as the largest and most important single item on our future development program."[65]

By "electronic computation," he meant electromechanical machines like those of Howard Aiken up the road at Harvard (although finally, some five years on in 1952, Aiken's Mark IV would be all electronic). And he assumed much if he took electromechanical computation to be "the largest and most important single item on our future development program." And he understood little if he felt that a machine like Whirlwind would postwar put MIT "in a much less favorable position." Such myopia makes some sense only in view of Ivan Getting's pronouncement that engineers of the day just didn't get it because of their limited training or complete absence of training in electronics. Rad Lab physicists and those engineers from the Servomechanisms Lab like Forrester and Everett, as well as other bright boys who spent time with electron tubes and radar pulse circuits, had inherited the new kingdom. What still remained on the horizon was their coronation.

Caldwell allowed a solid but non-Rockefeller opportunity to slip from his grasp when in the summer of 1945 a pre-ONR Navy extended the Center for Analysis an invitation to submit a proposal for an electronic computer.[66] Caldwell declined on the advice of the Rockefeller Foundation's Warren Weaver, a confidant of Caldwell's and a past grantor to the Center. Weaver's advice hinged on his determination that "the Navy would not be a good sponsor for the development which involved a major and basic scientific program," or so wrote Caldwell that September in explanation to Vannevar Bush. [67] It must have appeared to Caldwell that Weaver, as both director of the Rockefeller Foundation's Natural Sciences Division and as chairman of the Naval Research Advisory Committee, would have been privy to some inside information to have prompted such a recommendation. Or maybe the Navy was easy to decline if Caldwell was waiting on Bush to push through Congress his anticipated National Research Foundation, which could well have provided the Center with unlimited funding, if Truman had not vetoed it. Whatever the reason for rebuffing the Navy's generous offer, Caldwell seemed expectant of something in early 1946 when he wrote to Weaver about MIT having the "key men required for the theoretical, developmental, and engineering aspects of the problem."[68]

In 1946, the Rockefeller $100,000 grant came in, *not* to build a computer or even to produce a design for a computer "but rather [directed] at an appraisal and development of fundamental methods, both mathematical and instrumental, that would provide a well-founded basis for the subsequent design and construction of a machine."[69] Caldwell was not bothered a bit with any concern that the Center might be duplicating the bright boys' work, and he told Forrester so in a memo.[70] He was still beguiled with the prospect that MIT's resident math genius, Norbert Wiener, might be coming to the Center.[71] It didn't happen; Wiener politely showed no interest.

Caldwell may not have seen the end coming. It was easy to eye the new electronic, digital machines with incredulity and dismiss them. They were huge and ungainly looking, like a gymnasium stuffed to the rafters with thousands of glowing electron tubes—hot tubes that when massed together by the thousands created such heat as to need an air conditioning system nearly equal to the size of the machine itself. And because the tubes could blow out at any moment, every calculation was a huge gamble. Being so erratic, how then could such chancy beasts be relied upon to control the actions of yet another machine, even a simple thing like a cardpunch reader or printer? And what about the constant upkeep for all those electronics? Who would be daffy enough and deep-pocketed enough to endure the frightful expense of paying to maintain the unreliability of the Rube Goldberg-looking curiosity?

What were these kids getting everyone into? Weren't these digerati aware of the solid tradition and true-tested practice of analog computation, of analog servomechanisms, and the thousands of analog engineers working daily to push the technology to greater achievements? Were these bright boys blind to industry's multimillion dollar research laboratories, the country's engineering schools, and the countless equipment manufacturers where analog was the undisputed ruler of a domain that rejected all that was electronic and sniggered at the mere suggestion that the behemoths had a place in their world? What possessed the digerati to create that unnecessary extra step of transforming every instruction into binary digits? When a typewriter key is pushed, the key directly strikes a sheet of paper and makes its mark. Why then take such a direct and natural physical action and yoke it to unnatural binary digits, which then, in turn, need more binary digits to cause the typewriter key to strike the paper? Didn't these boys live in heated houses where there were thermostats on the wall? Couldn't they see the simple analog action of a thermostat maintaining control of a furnace to keep their homes warm? In the face of all that analog firepower, what in heck were these bright boys trying to prove? And was it worth it?

But what if all the doubts and misgivings could be erased? What then of the analog crowd and its fearful contempt of these electronic, digital upstarts? Forrester came through with marginal checking to locate and switch out electron and electrostatic tubes and other electronic components before they could do any harm. The five-digit multiplier enabled billions of reliable calculations without error. Electrical power was relatively cheap and air conditioning was on a postwar rise almost everywhere. And didn't it make good reductionist sense when doing a calculation or running any machine to reduce things to their simplest, like using a base two system over a base ten, where all digits—whole numbers, fractions, decimals, percentages, and degrees—and maybe also the English alphabet, were all reduced into combinations of just two binary digits? Sure, people couldn't read in binary, but machines could . . . and did it real fast. What if letters typed on a typewriter, numbers punched into a calculator, degrees on a thermostat, hours on a clock face, a speedometer, the percentage of someone's wages that went to the taxman, the vocal sounds from a telephone conversation, even the music of one's favorite orchestra, could be reduced to combinations of 0s and 1s? Wouldn't the thought of all that malleable commonality resident in just two digits spark someone's curiosity to do something with it? Would someone ever notice that what took separate and different analog devices to handle, like calculation, communication, and control, might well be harnessed by a single machine? Then too, maybe that single machine might be made to calculate, communicate, and control in real time? The bright boys thought that it was all worth it and persisted in the conviction that it could be done. By 1947, they had already invested three years of their lives in that belief.

If Sam Caldwell needed to toss any blame around for his fate, a goodly portion could well have been heaped upon the problems of the day, which were getting too large and too complex for the analog world to handle. An obvious one was that of the decennial US Census. The exponential increase in the growth of the country's population was straining Herman Hollerith's tabulating machines at the Bureau of the Census. The population had climbed from 63 million to 92 million to 132 million, and by 1950 was tipping toward the 152 million mark. The Baby Boom was cranking up (1946–1964), and waves of inner city GIs were escaping into suburbia in search of Levittown nirvanas for their families. All of which must have seemed like doomsday scenarios for the struggling Census Bureau.

It was reckoned that using the existing tabulating machinery, it would take nearly eight years to produce the hundreds of thousands of statistical tables and data necessary to properly administer the country. Before the 1950 Census was complete, the 1960 Census would be nearly ready to begin. It was an intolerable situation, especially for Congress, which used population shifts to determine the makeup of the House of Representatives. In 1946 the Census Bureau, members of which had attended the Moore School lectures, contacted Eckert and Mauchly to design a machine for their statistical purposes. In the following year Eckert-Mauchly produced the plans for UNIVAC, the Universal Automatic Computer, and in 1948, the Bureau of Standards and the Census Bureau ordered a machine for processing the 1950 census. The Census Bureau ordering an electronic digital computer like UNIVAC was an eye-opener for anyone who previously assumed that only the military—or maybe the Atomic Energy Commission—were appropriate customers.

Who would be next? The obvious targets were those companies that also used Census Bureau statistics—insurance companies and banks. Customers don't get much bigger than those heavy hitters. Government, insurance companies, and banks were some of IBM's core customers, which must have made Senior Watson choke on his remark about "a world market for maybe five computers."

The Census was a large obvious problem but only the tip of the iceberg in terms of the new and difficult problems facing business and industry, science and technology, as well as the military and government in the late 1940s. One of the biggest troublemakers to blame for many of the newest problems emerging in science, technology, and engineering was calculus. Calculus had first leapt out at the world when Gottfried Leibnitz published his mathematical breakthrough in 1684 (although Isaac Newton claimed that Leibnitz stole his idea of *fluxions*, calling it calculus). Calculus was the reason why people started calculating in earnest, and it is what spawned the need for computers.[72] The very first computers, however, weren't machines at all, but people. Individuals in the seventeenth century were called computers. Sometimes there were rooms full of computers sitting all day multiplying and dividing numbers for their boss.

Except for the substitution of a desk calculator for pen and ink, there was not much difference between the computers of 1684 and those 263 years later who worked on George Dantzig's 21-day problem in 1947. And just like Dantzig's employees, the calculations of early computers were most times related to a single problem. Sometimes the work carried on for years at a time; in the eighteenth century, Gaspard de Prony's trigonometric tables took six years to complete. A particularly tricky problem of the time was calculating the return of Haley's Comet, which we now know returns every 76 years. It was easy to predict the comet's return if historical records were taken into account, which for Haley's Comet stretch back to 240 B.C. However, to accurately calculate a huge elliptical orbit that takes decades to traverse was impossible without calculus. Since the comet moved through the universe and was affected by planets, moons, and other celestial bodies, its trajectory through space was difficult to figure. Enter calculus with its basic ideas on velocity, acceleration, height, slope, and situations involving rates of change from place to place or rates of change over time, and situations where all these variables interact. In short, calculus is the mathematics of motion and change. This new math was just the ticket for predicting the return of Haley's Comet and for astronomy in general.

It was also the ticket for blasting a bomber out of the sky with an anti-aircraft shell or for sending a missile 6,000 miles down range to a target. However, even with the magic of calculus, a roomful of human computers with their desktop calculators ablaze were easily outmatched by the problems they faced. At the Army's Aberdeen Proving Grounds human computers with calculators were assisted by a Bush Differential Analyzer until ENIAC was rolled in to take all their places. The great chain of human computing from Leibnitz to Aberdeen was over. It was not a newer, better, more advanced analog computer that rolled into Aberdeen or ran the numbers for the Census Bureau or freed Bud Baxter and his mates from their comptometers.

That an electronic digital computer was behind all these emancipations did not go unnoticed. A noose was slung over the analog world and began to draw close. And when the raging debates of quantum mechanics, begun in the 1930s, were added to the number-crunching fray, the analog noose grew ever tighter. Quantum mechanics didn't seek answers to comets zipping through the universe. Quantum mechanics posed questions about—and wanted answers to—the orbits of electrons around a nucleus or Erwin Schrödinger's equation for the electrical energy of the interaction between them.[73] A newer, smaller terra incognita began to open wide for exploration. It was a world on the atomic scale, of nuclear physics and elementary particles, where von Neumann dealt with things called BBB particles, which were 10^{27} or one billion times one billion times one billion.[74] This unimaginably small world, invisible in 1947, could be explored only by modeling it with mathematics . . . and a computer. It was the chance to explore the quantum universe, to gain knowledge about how it works, and to understand how it affects everything else, that sent von Neumann dashing from Cambridge to Philadelphia. And if that was what the trendsetter was doing, what of every other mathematician who followed his every movement? They too scurried around, each searching for his very own TARDIS to take on voyages of discovery. There existed in 1947 barely a handful of electronic digital computers and, except ENIAC, all were either in early stages of construction or, like Whirlwind, only on paper.

Very early on in the rise of electronic digital computers, analog computation seemed willing to concede the field of high-level numerical calculation. Even electromechanical leviathans like Howard Aiken's Mark I, Mark II, and Mark III and IBM's Selective Sequence Electronic Calculator (SSEC) would ultimately falter and stumble: the Mark I was disassembled, the Mark II and Mark III were shipped off to the Naval Proving Ground at Dahlgren, Virginia in 1948 and 1950, respectively; and the SSEC left Manhattan in pieces in 1952 for parts unknown and was quickly replaced by the all-electronic IBM 701. Calculations that went way out to the right of the decimal point were no-man's land for these behemoths. Wisely, they shied away from high-level matters, as if retreating from anything having to do with the cerebral cortex, preferring a more familiar and more comfortable place like the ancient limbic system, a world of control. In his paper, "A Great Disappearing Act: The Electronic Analog Computer," Chris Bissell follows the beasts, both electromechanical and the wheel and disk integrators, into their special world of control.[75] They were originally created to serve that world and it was there that they were giants: fast, direct, single-purpose machines that controlled other machines and processes, machines that modeled and simulated dynamics and control principles. For example, they were valuable in chemical plants where they monitored temperatures, pressures, and flow rates, and then sent corresponding voltages to various control devices, which, in turn, adjusted the chemical processing conditions to their proper levels. Others were quite simple ones like an automobile speedometer that measured the rotations per minute of a car's drive shaft and then translated those rotations into measurements that displayed on a dashboard as miles per hour.

Hundreds of thousands of analog machines controlled a world where hundreds of thousands of analog engineers and technicians were necessary to keep them all humming. As important as control computing was, ENIAC was built to be above all that. But not so Whirlwind. Whirlwind's parents intended it to be a control computer. Whirlwind would be designed not only to crank out some high-level calculations for the scientific types, but, more specifically, to also dive deep into the world of control. Control was where Valley said the analog computer-servo crowd lived, and it was there that they would fight like crazy for their homeland.

Coming of Age

In February of 1947, Warren Weaver paid Forrester a two-hour visit. Weaver came not as the director of the Rockefeller Foundation's Natural Sciences Division but rather as chairman of the Naval Research Advisory Committee in order to check up on the ONR's machine. By checking up, he would mean: what are you doing with the Navy's money and what's taking so long? They seemed like fair enough questions. The Navy had an investment and Weaver wanted to know how that investment was being managed.[76] Weaver was not satisfied with the answers, and questioned if Forrester was very sure of himself and his machine. Was the development thus far the expected flight analyzer or was it now, as Forrester indicated, headed in the direction of a fire-control computer? Weaver reported that Forrester and Everett did not know if they were making a machine that was "biscuits or cake."[77] It was an odd remark to come from the 53-year-old former mathematics professor on an official ONR visit. But then again, Weaver was a big fan of the inverted logic in Lewis Carroll's humor and his comment on Forrester's unsure responses was typical Red Queen.

Weaver supposedly owned the largest collection in the world of foreign language translations of *Alice in Wonderland*; Margaret Meade claimed to have personally translated it into Swahili for him. Weaver arrived at MIT knowing full well the contents of Forrester's letter to Lt. Commander Knutson, but did not let on during his inspection tour.[78] Forrester must have felt a bit insecure with Weaver's opinion of the project and of his leadership, for shortly thereafter he and Everett started putting together L-1 and L-2 as a definitive response. The reports would fail to allay Weaver's skepticism about Whirlwind and the bright boys.

Warren Weaver was an odd character in the early years of electronic digital computing. His backstory meanders throughout the mathematics world of the 1940s, especially at those points where it bumped into computing. New York University would honor his association with computers enough to name its computer science building Warren Weaver Hall.

Although not a creative or original mathematical thinker, Weaver had talent for attaching himself somehow to others' more innovative output, and as such, lived vicariously at the epicenter of some great moments. Many of his mathematical relationships came courtesy of his position as director of the Rockefeller Foundation's Natural Sciences Division (1932–1957), a directorship bestowed upon him by Max Mason, then president of the Foundation as well as an old colleague and friend from their days together at the University of Wisconsin.

Weaver is the guy who encouraged von Neumann to author the EDVAC Report of 1945 while the latter was staying at Los Alamos and had some time on his hands. His alliance with Sam Caldwell, the Center for Analysis, and his affinity for things analog helped to keep afloat the false hope that the Center might somehow revive itself. With Sam Caldwell, Weaver tried to lure Norbert Wiener to the Center. Barely a month after his meeting with Forrester, Weaver, on March 4th wrote to Wiener, feeling out the great MIT mathematician with, "I have wondered if it were unthinkable to design a computer which would translate?"[79] He had gotten to know Wiener through his wartime job with the OSRD, which was doled out to him by the Center's cofounder, Vannevar Bush. Weaver, in turn, doled out wartime contracts to Wiener and his partner, Julian Bigelow. Wiener claimed in *Cybernetics* that Weaver had, in 1944, published his and Julian Bigelow's secret wartime work "on predictors and wave filters,"[80] which later became a large part of Shannon's mathematical predictor theory in "A Mathematical Theory of Communication."[81] As possible recompense, the Rockefeller Foundation in 1946 (through Weaver's office) provided Wiener with a substantial grant to "pursue research on the physiological and mathematical foundations of cybernetics."[82]

Later in the war, Weaver headed the Applied Mathematics Panel where his wartime assistant was Mina Rees, who would eventually run the ONR's Mathematics Branch beginning in August of 1947 and have direct authority over Whirlwind and its spiraling budget. In 1949 when crunch time came for the bright boys, Mina Rees was the one with her fist around the Navy's wallet. In 1948 Claude Shannon published his information theory, "A Mathematical Theory of Communication," in *The Bell System Technical Journal*, and it was Weaver's philosophical essay on the theory that shared space with Shannon in his coming-out volume published in 1949 by the University of Illinois Press. There are more than a few scholarly citations that still credit Weaver as coauthor of Shannon's information theory. When Weaver met Forrester in February, he had already granted Caldwell his $100,000 to "study" computing technologies, and Caldwell had already sent Forrester a note to the effect that his Center was going to compete directly with the bright boys, writing that it was, "the price which must be paid for independence of thought and action."[83]

Weaver's own area of interest was machine translation or MT, which he pursued with great vigor throughout the late 1940s and early 1950s. During that period, he "started thinking and talking about the possibility of using computer-like machinery for carrying out part or all of the operations involved in translating from one language to another."[84] "Computer-like machinery" was further fleshed out by a Weaver staffer, John Marshall, in a letter to the British literary critic Ivor Richards, who was then teaching at Harvard. Marshall explained how Weaver was tinkering with "the possibility of utilizing the new super calculating machines in translation."[85] Weaver's own colleagues at the Rockefeller Foundation circulated notes questioning his computer translation idea, which they considered, at best, overly fanciful.[86]

On March 6, 1947, Weaver met in the United States with Andrew Booth from Birkbeck College at the University of London. The previous year, Booth had toured the Moore School's computer facilities; and the Rockefeller Foundation paid for the tour. Booth was looking for financing in order to build a computer (an electronic digital one) at the University of London. Weaver indicated, "Americans might be interested in new non-numerical uses for computers, and suggested translation as an idea for Booth to think about."[87] "Non-numerical" was the Red Queen's way of saying "control computer," with the computer controlling a process to pull translations out of a database. Control computer was what Forrester said he intended for Whirlwind. The Navy was interested in "pure and applied mathematics, statistics, and computer development with its related numerical analysis."[88] In other words, numerical machines. Did Weaver want biscuits or cake?

Wasn't this, looming up before them, a perfect opportunity for all to try to work together? The timing couldn't have been better for some sort of new-tech rapprochement to have taken place among Forrester, Caldwell, and Weaver—something in those earliest days of electronic computing that would be mutually beneficial to the Navy, MIT, and the Rockefeller Foundation. It might well have saved a lot of time, money, and anguish if all three collectively pioneered essentially differing aspects of the same new technology. Such mutual efforts would have moved the technology forward much faster, and maybe even hurried along the eagerly awaited return of MIT's world-class reputation for Caldwell who desired it most. It never came to pass.

In reality, Weaver knew, as everyone else was quite aware, that Caldwell was an immobile fixture in the ancient and honorable society of analog computer builders. Building an electronic digital computer was out of his league, although Weaver had given him $100,000 in seed money for some kind of as-yet-unspecified but electronic machine; and it was also well beyond Norbert Wiener, who was a theoretician and not a computer builder.

On April 30, 1947, Wiener diplomatically but firmly declined Weaver's hint at joining forces. Wiener conceded that in the "mechanization of speech" more could be done but "I am frankly afraid the boundaries of words in different languages are too vague and the emotional and international connotations are too extensive to make any quasi mechanical translation scheme very hopeful."[89] Weaver would try Wiener again that May, but once again he was gently rebuffed. With his wartime aide-de-camp Mina Rees having "won the title of mathematical architect of the ONR,"[90] might Weaver then have looked toward the ONR for MT help? The Pentagon, after all, was very interested in translating Russian to English; the Navy had ERA working on cryptologic devices for cracking the Russian VENONA code, including the assembly of the Atlas computer to aid in that cryptological pursuit.

Soon after Wiener's final refusal to Weaver, Caldwell seems to have quickly slipped from view. A year later in *American Scientist*, Weaver was touting the mantra of electronic digital computing, comparing it to more like "a human brain than like the traditional mechanical computing device of the past."[91] If Sam Caldwell had read the piece, the word "past" might well have been a sad one for him to ponder:

> They have memories in which vast amounts of information can be stored. They can be "told" to carry out computations of very intricate complexity, and can be left unattended while they go forward automatically with their task. The astounding speed with which they proceed is illustrated by the fact that one small part of such a machine, if set to multiplying two ten-digit numbers, can perform such multiplications some 40,000 times faster than a human operator can say 'Jack Robinson.' This combination of flexibility, capacity, and speed makes it seem likely that such devices will have a tremendous impact on science. They will make it possible to deal with problems which previously were too complicated, and, more importantly, they will justify and inspire the development of new methods of analysis applicable to these new problems of organized complexity.[92]

What a difference a year makes.

When the analog gear works was finally disassembled and yanked out of the Center, Weaver eulogized it to Caldwell:

> It seems rather a pity not to have around such a place as MIT a really impressive Analogue computer; for there is vividness and directness of meaning of the electrical and mechanical processes involved . . . which can hardly fail, I would think, to have a very considerable educational value. A Digital Electronic computer is bound to be a somewhat abstract affair, in which the actual computational processes are fairly deeply submerged.[93]

The Center for Analysis would later reemerge as the Center for Machine Computation in 1957, under the direction of physicist Philip Morse.

Maybe Weaver was just being kind to an old friend, maybe not. Weaver was on a mission with its own personal agenda—an impossible one for 1947. And for the bright boys: beware a man on an impossible mission. The problems Forrester encountered trying to yoke a computer to a flight analyzer paled, and grossly so, in comparison with the impossibilities that Weaver faced in the 1940s with machine translation of human language. Norbert Wiener was right in being afraid that "the boundaries of words in different languages are too vague." They were then and would elusively continue to be so up into the 1990s. The closest MT would come to anything substantive in any reasonable time frame was the January 7, 1954 public demonstration of a Russian–English machine translation system at IBM's headquarters in Manhattan. It was a small-scale experiment of just 250 words and six grammar rules,[94] and took place seven years after Weaver's first visit with Forrester.

It was not until the 1960s that a computer at Harvard could parse words in English, and even then only imperfectly so. For Weaver, MT was akin to cryptography: all that needed to be done was to break the code of one language to reveal the target language beneath. From his viewpoint, a sentence in Chinese was really English but written in code called Chinese; crack the code and there's your English. This was not so simple a task, and he eventually got the picture.

Weaver's reign over Whirlwind was filled with complaints that it was "mathematically unsound . . . and grossly overcomplicated technically;"[95] that it was just an overly expensive knockoff of von Neumann's IAS machine[96] . . . and was of little value for scientific research.[97] However, even a cursory assessment of that reign exposed enough conflict of interest as to make any judgment on Whirlwind highly suspect. With all his ONR and Rockefeller ties and interests, with the power of money and influence he wielded, with his cronies and confidants seemingly everywhere, and with his own personal agenda and pursuits dripping from the computer projects with which he came in contact, Weaver could not be the paragon that he should have been. He had too many hats in the ring. His actions showed a person with sufficient conflict of interest such that everyone might have been better served had he recused himself from any dealings with the bright boys.

The bright boys were at risk. Relative unknowns like Forrester and Everett, as Mauchly and Eckert before them, were vulnerable to predatory academics who might resort to influence instead of innovation to get ahead. Weaver's contact with the bright boys should have been solely on behalf of the Navy and taxpayer dollars. It appeared otherwise. From August 1947 onwards, Whirlwind's fate would largely rest in the hands of Warren Weaver's old friend, Mina Rees, who remarked sight unseen about Whirlwind and very White Queenly of Weaver's report, "too much talk and not enough machine."[98]

The Barta Building

The ornate escutcheons on the Barta Building, each inscribed with a large floral "B," are separated like belt loops around the girth of the two-story structure. Rusticated concrete windows and corner trim set off and brightly contrast with the dark exterior of red brick. It is said that the beaux-arts style of the Chicago World's Fair of 1893 inspired the building's design. A grand cartouche etched with the year 1904 is perched majestically above the massive oak door at 211 Massachusetts Avenue. What a wonderful place for a computer to grow up. And in the spring of 1948, the bright boys began to raise their offspring there.

The Barta Building was two blocks from MIT's front door, just around the corner from John Marchetti's Air Force Cambridge Research Laboratory at 224 Albany Street, and a quick jaunt from George Valley's office in the Physics Department. Cambridge city records say that E&R Laundry, an industrial cleaner, put the building up. That's probably why it had a 60-foot smokestack out back. Still there but in disuse when the boys arrived, the smokestack's massive base is all that remains today; its ancient iron hatch inscribed in Latin boasts that the furnace was made in New York City at the turn of the century. The floral "Bs" came later with the building's second owner, the printer Barta Press, which sold it to MIT. A bit grimy and worse for the wear after sitting on the sidewalk for 44 years, the building's original sheen had long since tarnished by 1948. But it still had charm. Any bright boy needing to get away from the others for a bit of seclusion could nip off to the charming, rooftop turret with its copper cupola. Beneath the turret a fancy trim of ornamental molding ringed the second floor; molding interspersed by small gargoyles with faces like theatrical masks . . . some smiling, some frowning, some angry, some crying, some laughing, and some aghast—every human emotion at the ready—to bear witness to whatever performance the boys had in store.

Exterior of Barta Building, 211 Massachusetts Avenue, cicra 1949. (Photo courtesy of MITRE Corporation.)

From February onward, time schedules were drawn up and the layout of the computer began. The bright boys kept monthly Summary Reports that kept the ONR posted on every aspect of their progress. Forrester intended that the reports be so thorough and exacting that they would make a "virtue of their frankness" such that the ONRv would never call into question Whirlwind's forward progress. They had the opposite effect. The reports showed plainly how the bright boys went about things: too thorough, too painstaking, and too expensive in the ONR's estimation.[99] Things backfired even worse when the now overly curious ONR sent agents to check on the activities in the Barta Building. With what must have seemed something akin to the frequency that Queen Elizabeth took in sending voyages to the New World, Mina Rees sent expeditions to Cambridge.

First to arrive after the Weaver inspection tour was the Columbia mathematician Francis Murray, who came away with generally positive marks for the bright boys and their machine, but suggested Whirlwind needed a "thorough mathematical analysis."[100] Next came the von Neumann expedition that lasted three days and ended with kisses for Norman Taylor. He was followed by the duo of John Curtiss and Harry Huskey, both from the National Bureau of Standards or NBS, a huge recipient over the years of ONR munificence.[101] They struck gold, not for the queen, but rather for themselves and the NBS. In 1947, Curtiss had founded the National Applied Mathematics Laboratories at the NBS and was in the midst of building its own computer called the Standards Eastern Automatic Computer or, as it came to be called, the SEAC. The Barta Building proved to be a trove of technical wisdom for his organization. Curtiss had also been a presenter and participant at the Moore School lectures and was predisposed to the coming of electronic digital computers. Whirlwind was, in his judgment, good enough to copy, and that's just what he did. Harry Huskey was one of the engineers who had wired up ENIAC. Then in August of 1946 he went to England for a year at the National Physical Laboratory to work with Alan Turing on wiring the Automatic Computing Engine (ACE). Returning in 1948, he had just joined the NBS, and together with Curtiss was about to begin designing SEAC's brother, the Standards Western Automatic Computer (SWAC) at the Institute of Numerical Analysis in Los Angeles, where Curtiss would become the director. SWAC's arithmetic unit would come into the world wearing a circuit design courtesy of Whirlwind.[102]

Ironically, one of the first jobs that Huskey planned for SWAC was German to English translation. Of course, there may well have been less irony there than meets the eye. In May of 1949 the *New York Times* reported on Huskey's demonstration: "While the exact scope that the machine will have in the translating field has not been decided, the scientists working on it say it would be quite possible to make it encompass the 60,000 words of the Webster Collegiate Dictionary with equivalents for each word in as many as three foreign languages."[103] This first newspaper article ever on the new field of machine translation, by no less esteemed a news organization as the *Times,* gave legitimacy to Weaver's quest. It would be publicity beneficial to MT earned indirectly courtesy of Whirlwind's cloned circuits. If Forrester's monthly reports had backfired into unwanted scrutiny, the visits themselves turned out to be promotional coups. The bright boys and their machine were winning over people who they were not supposed to win over.

By September of 1948 even MIT began to take notice. The bright boys were converting mathematical, logical, and abstract concepts into actual working machinery.[104] Trucks were arriving from Sylvania with the racks that formed the skeletal outline of the computer's superstructure. Clusters of assemblies and subassemblies of components and subcomponents began to flesh out Whirlwind's bones. MIT leadership "became aware that they had a partly finished, well-begun computer of unique design on their hands."[105] The behemoth was rising.

The bright boys would design, build, test, integrate, redesign, rebuild, retest, and reintegrate.[106] They felt unbound from any set of engineering requirements, even their own. If along the way they hit upon a better process or component, they jumped at it, tested it, and then, if it improved their design, they integrated it into their machine. Such an unexpected discovery came about when dealing with the pesky, ever-unreliable electron (vacuum) tubes of the day. They happened upon the problem that caused the tube's short lifespan. The source was the fabrication of the nickel cathode within the tube itself: it was the tube's own innards that doomed it. The manufacturer, in order to make the cathodes easier to shape, included a small amount of silicon in the nickel from which the cathode was made. After some 500 hours of operation, a high resistance, monomolecular layer of silicon would migrate to the surface of the cathode and bias off the flow of electrons. The problem had not been, as assumed, a failure of the cathode coating to emit electrons, but instead was internal resistance to the flow of current. The bright boys had tubes built to their own specifications of high-purity nickel cathodes even though they were somewhat harder to shape. The result increased the life of vacuum tubes a thousandfold. Because of their efforts Sylvania produced the 7AK7 tube in 1948, which into the late 1950s was the standard tube for all computers.

Such an eye-opening lesson in the bright boys' brand of quality control had a profound reaction at Sylvania. "The Sylvania engineer who was in charge of making our tubes," recalled Forrester, "realized that production should be insulated from the past practices of making cheap tubes. He insisted that the tubes be made in a city where tubes had never been made before so that bad practices did not have to be unlearned." As Forrester concluded, "Very insightful."

Such improvements escaped the ONR's purview and went down simply as line-item expenses—irritatingly intolerable expenses. With the 7AK7 undoubtedly an integral and expensive part of many, maybe all, of the other ONR-sponsored computing projects, a little thanks might have been more appropriate. Mina Rees made reports to the new Chief of Naval Research, Admiral T. A. Solberg, who, in turn, approached MIT's president Karl Compton on September 2, 1948 with the news that Project Whirlwind's funds "future commitments and rate of expenditure be scaled down" until such time as both the technical and financial requirements of the project could be reevaluated.[107] It came at a bad time for Compton, because he was getting ready to resign his MIT presidency and go to Washington to replace Vannevar Bush as head of the Research and Development Board (RDB), which had temporarily replaced the OSRD until such time as the National Research Foundation would come into being (initially vetoed by Truman in 1947; later signed into law by Truman as the National Science Foundation in 1950).

On September 8th, Compton asked Nat Sage to get him up to speed on Whirlwind, and on electronic digital computers generally. He needed the information for two reasons. First, he knew little about the new electronic machines, and second, he would be in a position in Washington to do a little PR for Whirlwind if he had a good document to circulate around.[108]

Nine days later, on September 17th, Compton had his report. Forrester and Everett, together with Hugh Boyd, Harris Fahnestock, and Robert Nelson, put out Servomechanisms Memorandum L-3, titled "Forecast for Military Systems using Electronic Digital Computers." L-3 had two parts: a budget proposal covering fiscal years 1949 to 1953[109] and a two-by-three foot foldout page on which there was a grid with a 15-year forecast (1948–1963) of computers in the military, covering guided missile data reduction, cryptography, high-speed computer networks, interception networks, air-traffic control, industrial process control, simulation and training, and logistics. Each column in the grid represented a year and each row was dedicated to a future military capability that computers would serve.[110] L-3 looked like a graphical companion to L-1 and L-2. The practical engineering side of the bright boys could not resist making abundantly clear the potential applications for their new electronic digital wonder machine—some 15 years into the future! On a broader level the report represented a first-ever call to action for the country to begin a computer development program on a national scale; a claim that it would be a far better alternative than the scattered investments in one-off machines as was the case up to 1948.[111]

For a national economy invested up to its chin in analog machines, that would be a hard sell. However, the military was quite a different story. The report seemed to be a perfect fit. L-3 could have slid nicely between the covers of von Karman's *Toward New Horizons*. And much like *Toward New Horizons*, it too was an eye opener. The bright boys pointed computing toward a future that was compelling and undeniable. And what L-3 said about its authors was equally remarkable. It displayed an amazing range of experience, wisdom, and maturity in all things electronic and digital acquired by the bright boys since 1946. They had come far.

However, technology and national interest aside, in the end it all came down to money. MIT met with the ONR on September 22nd about the impending financial crunch for Whirlwind for fiscal year 1949, which would end on June 30, 1949. Forrester's request for $150,000 per month ($1,800,000 for the year) was met by the ONR's firm counter of $900,000 (or $75,000 per month).[112] A draconian cutback for Whirlwind. The machine's monthly glass bill alone would eat up a huge chunk of that. Whirlwind's 32 electrostatic memory tubes were costing $32,000 a month. Whirlwind was big and expensive, and none of it was news to anyone, especially the ONR. The computer was building out to cover an incredible 2,500 square feet of floor space—two and a half times larger than ENIAC. The ONR offer was not killing the beast outright; it was squeezing it into extinction by the month. Nat Sage offered a way out, saying that $1,200,000 would do the trick for fiscal year 1949, and that a carryover from fiscal year 1948 would up the ante to about $132,000 monthly in working capital for the upcoming year.[113] Relenting a bit during the meeting, the ONR conceded that more money would be forthcoming if Whirlwind seemed to be making and keeping a good faith effort toward the $900,000. Flamm suggests in *Creating the Computer: Government, Industry, and High Technology* that there were other motives afoot at the meeting, unstated ones like the ONR's sensitivity about the "large fraction of public research funds for computers going to MIT."[114]

Then too, was the ONR ever going to see MIT kick in a few bucks toward Whirlwind's monthly meal ticket? Whirlwind stopped having a Navy mission when the flight analyzer was dropped, yet the ONR was still paying top dollar as though the mission was still ongoing. Whirlwind was the only electronic digital computer that MIT had on campus and would undoubtedly go to enriching the educations of thousands of MIT students. Did any of that digital advantage count toward any financial consideration on MIT's part? Maybe like offering to split the difference in the contested $75,000 per month? Nothing of the sort was forthcoming from MIT. With some 200 universities participating in the ONR's work programs, no university or college would be silly enough to suggest sharing expenses. That could well turn out to be a dangerous precedent. Besides, the ONR put itself out as the Daddy Warbucks of research, funding 40 percent of all basic research in the United States and supporting nearly half of all doctoral students in the physical sciences.[115] In support of computing, the ONR's share was almost total, and would continue to be right to the late 1950s.[116] If the ONR cut anyone off from funds or squeezed a project, there was no recourse. You took the hit and moved on to something else. Building a computer under the ONR's auspices was precarious. The bright boys got their $1,200,000 for fiscal year 1949, but for fiscal year 1950, their budget was slashed way back to $750,000.[117] That gave them nearly a year, until June 30, 1949, to come up with a plan for their deliverance.

Chapter Six

All Together Now

*An invasion of armies can be resisted, but not
an idea whose time has come.*

—Victor Hugo

By 1949 America was reverting back to the land of plenty.
There was more of everything, and it showed. Hemlines went
down to mid-calf and stayed there.

Over $40 million in war matériel was declared surplus and
fed back into the system. Before long military surplus stores
sprouted up everywhere and did a brisk business. The seam-
stress of the house could buy an aviator's parachute for $13.95,
which yielded 45 yards of fabric for skirts, blouses, lingerie,
and baby clothes. Perfect around the farm for brooders or
milking machines were government surplus diesels, ranging
from $178 to $480. Some began a taste for four-wheel drive
vehicles with a like-new Army surplus Jeep. An enterprising
merchant in Quakertown, Pennsylvania ran a newspaper ad
for military flamethrowers, pitching the weapon as a handy
household gadget that "destroys weeds, tree stumps, splits
rocks, disinfects, irrigates. 100 practical uses. $22 for 4 gal.
tank, torch, hose." Jay Forrester had an eye for war surplus as
well. One April evening in 1949 while leafing through a copy
of *Electrical Engineering* he spotted an intriguing advertisement by the Arnold Engineering Company for a substance
called Deltamax,[1] which would eventually lead him to one of
the greatest discoveries in the history of computing.

Deltamax was a magnetic substance, a "specially treated
nickel-iron alloy . . . developed by the Germans and used during World War II in naval fire control equipment."[2] Deltamax
(known in Germany as Permanorm 5000-Z) was German
invented and patented and, along with 50 million other

German patents, was plundered by the Technical Industrial Intelligence Committee, whose job it was "to look into every segment of the German industrial landscape and obtain any information that might be of interest to American companies."[3] Every "segment" included 3.5 billion pages of documents, 300,000 pounds of equipment and product samples, and 200 tons of other materials.[4] Technically owned by the US Alien Property Custodian's Office, the patent booty was routinely passed through to the commercial marketplace for exploitation, potentially raking in billions of dollars for the new patent owners. John Gimbel's *Science, Technology, and Reparations: Exploitation and Plunder in Postwar Germany* puts the plunder-to-dollars conversion at a conservative $10 billion.

Plundered magnetic materials were hot items in 1949, if you could figure out what to do with them. ERA used captured magnetic drums on its computers, mastering the technology so well that IBM in early 1950 sought out a partnership to share in the expertise.[5] Bing Crosby used a captured Magnetophon tape machine to record his radio show, and then helped to hurtle the Ampex Corporation into the new industry of tape recorders and its resulting landslide in profits. As tape recorders and magnetic drums, these new materials and processes were breakthroughs at capturing and storing sound and computer data for high fidelity playback or secure data storage. Might Deltamax somehow or other, mused Forrester, staring at the advertisement, replace the memory storage in Whirlwind's expensively unstable electrostatic tubes?[6] "The idea immediately began to dominate my thinking, and for the next two evenings I went out after dinner and walked the streets in the dark thinking about it, turning over various configurations and control methods in my mind until I began to see a configuration in which we could use a magnetic element . . . "[7]

Something had to dominate his thinking other than the fast approach of the new fiscal year on July 1 when Whirlwind would be cut back to $750,000 through to the following June 30, 1950. If something wasn't done fast, memory tube costs would sink them all. 1949 was stacking up to be a gruesome year. The previous fall, Forrester and Everett had fought off the ONR's

Schematic drawing of Whirlwind computer layout.
(Image courtesy of MIT Archives.)

insistence that Whirlwind and von Neumann's IAS computer were mostly identical machines. So why should the ONR fund both? It was an observation from Francis Murray's report that Mina Rees passed to von Neumann for comment. Von Neumann didn't agree with Murray. Nevertheless, the bright boys had to rush about in October of 1948 to produce for the ONR "Comparison Between the Computer Problems at the Institute for Advanced Study and the MIT Servomechanisms Laboratory." Mina Rees put together her own expedition in the spring of 1949 accompanied by her own set of experts: Harry Nyquist from Bell Labs; Karl Spangenburg, head of the ONR's Electronics Branch; and Charles V. L. Smith, head of the ONR's Computer Branch. It was a so-so meeting with a generally favorable result, but without the promise of additional funds anytime soon.

In July of 1949 the Research and Development Board put together an Ad Hoc Panel on Electronic Digital Computers—of which Harry Nyquist was a member—to decide which of the 13 government-financed machines should be axed because of limited funds or lack of a compelling mission.[8] Its preliminary report in December eliminated Whirlwind. The Panel determined that the machine was too expensive and that no one had a use for it.[9] Forrester's rejoinder that ERA's program was three times that of Whirlwind was dismissed. Also that December, after the ONR's director Captain J. B. Pearson asked Forrester to project future costs, which Forrester did with a 1951 estimate of $1,150,000, a major kerfuffle erupted. First came the ire of R. J. Bergemann, Physical Scientist for the ONR's Boston Branch Office, complaining that the ONR was not getting its money's worth out of Whirlwind, and that Raytheon's Project Hurricane Computer (later known as RAYDAC) was technically superior.[10] Bergemann was quickly followed by Charles V. L. Smith, who found Forrester's refusal to recognize that there were no available funds "appalling" and found that Whirlwind's program was "excessive."[11] Mina Rees and Smith propounded Whirlwind's 1951 budget at $250,000 with a little wiggle room to no more than $300,000. Worse yet, the Navy indicated that it was ending its commitment to Whirlwind after fiscal year 1951.[12] Rees would personally journey to Cambridge to officially deliver the news. Entreaties for help from Forrester to Provost Julius Stratton and others in MIT management got no better than lip service.[13]

A few slivers of light penetrated the gloom in the Barta Building. The bright boys pulled off a small $216,000 contract ($9,000 per month for 24 months) with the Air Force's Watson Laboratories for a preliminary study of digital computer applications to air traffic control.[14] The project was supervised by Gordon Welchman together with four other bright boys (Harris Fahnestock, C. Robert Wieser, William Linvill, and Theodore Hildebrandt)

Interior of Barta Building with Whirlwind computer, 1951. (Photo courtesy of MITRE Corporation.)

plus Whirlwind's new resident mathematician Philip Franklin. (Forrester had bowed to Murray and von Neumann's suggestion to get a project mathematician.) I. J. Gabelman, program manager from Watson Labs, had selected the bright boys and their machine over the corporate giant Raytheon. A small victory in the face of a season of bad news.

Another glimmer of hope came from none other than Perry Crawford during a meeting with Forrester in Washington in early January of 1949. The Research and Development Board's Ad Hoc Committee on Air Defense, of which Crawford was a member, was mulling several Air Force requests associated with their air defense needs. With some eight different problem areas all relating to electronics, Crawford felt sure that some or all of it might bode well for Forrester. Basically, the Air Force realized that human identification and plotting of jet aircraft were too slow.[15] The Air Force needed what they vaguely referred to as an "automatic interceptor Director System," which was an electronic system that could somehow automatically vector jet fighters to their targets. Crawford encouraged his friend, saying that he would keep an eye out for future developments that could be promising to Whirlwind. In December, after Norman Taylor returned from a visit with the Air Force at Wright Field, Forrester heard yet again about the plight of the Air Force's "air intercept" dilemma. It seemed that jet pilots, because of the speed of their aircraft and their lack of adequate jet training, were finding it difficult to turn and come up behind enemy planes. Radio ground control and radar were proving inadequate to the challenge. "These people feel," reported Taylor, "if we can solve this one the Air Force has almost unlimited funds available. They are spending seven-figure money now and not improving techniques."[16]

The previous May, John Marchetti from the AFCRL on Albany Street had traipsed into the Barta Building accompanied by his colleague Lowell Hollingsworth. They had a scanning and tracking problem to work out, namely, the tracking of 300 aircraft using SCR-584 radar with M-9 directors controlling antiaircraft gun batteries.[17] It was quite similar to Marchetti's 1937 experiment at Fort Monmouth when his Army Signal Corps SCR-268 radar unit succeeded at detecting a B-10 bomber 12 miles out to sea, but this time the detecting was on a massive scale. Twelve years, World War II, the Chain Home, microwave radar, and the Rad Lab had intervened to present a new world to him where it was routine to theorize about radar detection of aircraft. Pulling it off for hundreds of aircraft at once was another story, but Marchetti and Hollingsworth unexpectedly stumbled across a guy who well understood their need. To Forrester it sounded all too familiar because he and Everett had addressed a similar situation in their L-2 report from 1947. Their neighbors from the AFCRL "seemed skeptical that digital computation could accomplish quite efficiently some of their jobs in ground controlled interception" but before they left the building, "their interest in digital computation seemed to liven greatly."[18] There was, however, a distinct and recurring kinship among Crawford, Talyor, and Marchetti's stories. All seemed to be converging toward the same place. Each story revolved around quickly acquiring information about incoming enemy air traffic, tracking the aircraft, and then sorting out that tracking information so as to direct interceptor aircraft on the shortest, quickest path to interdict them. The Soviet A-bomb explosion in August would undoubtedly add impetus to the convergence. For Forrester and Everett it appeared that at long last the rest of the world was finally catching up to their ideas on computer-aided command and control.

Auguries of impending doom mingling with the tantalizing potential for success flitted back and forth during the whole of 1949. Not knowing which would strike first, they forged on, building out their machine, stretching it out on either side of a long corridor, eight racks deep and 4,000 electron tubes removed from its central control room. A rug ready to be pulled out at any time from under their well-worked shoes was always at the back of their minds. Forrester penned in his Computation Notebook that Whirlwind's future seemed less than promising.[19]

In the face of such uncertainty, to keep a clear head while reflecting over an advertisement for Deltamax, debating with himself if and how the alloy could work as a substitute for electrostatic memory, was remarkable. He somehow kept his wits about him and blocked out the unsavory distractions ready to be dished out in the new fiscal year. He filled his notebook from the spring of 1949 with clear-headed, incisive notes regarding the exact three-dimensional configuration and the exact x,y,z-axis control voltages that would become the magnetic-core memory design that would dominate computer memory through to the 1970s.[20] A revised drawing in his notebook from October 9, 1949 shows a stack of cores that was very nearly his famous final design.[21]

The Nebraska boy who electrified screen doors and brought electric light to his farm, and the engineer who was known to his contemporaries as aloof, brilliant, and imperious both came together over that ad for Deltamax. Like the Splendid Splinter smashing the big one into the bleachers at Fenway, Forrester did likewise: with a single idea he reinvigorated the whole of computing. As George Valley remarked about the achievement: "I do not think anyone else would have developed core storage, had Forrester not done it. The people in other organizations who said they were doing it seemed too dilatory. The history of computer memory would probably have been that of transistor storage followed directly after storage tubes."[22] This view of the bright boys—these sons of Pasteur's "chance favoring the mind that is prepared"—is where the ONR, Warren Weaver, Mina Rees, MIT management, analog antagonists, and other Whirlwind doubters and naysayers should have directed some of their time and consideration.

Innovating as they did such critical new technological advances as real-time computing, parallel processing, reliability, modular machine design, electron tube manufacturing, electrostatic tube construction, and magnetic-core memory—all pulled off in an incredibly compressed period of time—is without parallel in the history of computing. And in 1949, the bright boys were not even half done yet with their string of innovation. A close look at what was really happening in the Barta Building or a quick glance through their research notes would have quickly differentiated Whirlwind from the IAS machine . . . or any other machine anywhere. It would have made the offer of $250,000 for fiscal year 1951 as preposterous to the ONR as it was to the bright boys.

Forrester wasn't the only one who walked around dreaming about magnetic- core memory. Twenty-year-old An Wang (Chinese-born Harvard research fellow, who later founded Wang Laboratories) had his magnetic epiphany while walking through Harvard Yard.[23] A recent Harvard PhD in applied physics, Wang started working for Howard Aiken's Mark IV in May of 1948 (a year prior to Forrester spotting the Deltamax ad), and was given the job of finding a way "to record and read magnetically stored information without mechanical motion."[24] Wang's June 29, 1948 notebook entry talks about storing data in donut shaped magnetic cores, then reading the data from and rewriting data back into them.[25] He found a Navy publication that described the characteristics of Permanorm 5000-Z, liked what he read, and then hunted down Deltamax. Wang's revolutionary idea

Interior of Barta Building at Whirlwind control room, circa 1950: (standing left) Jay Forrester and Bob Everett; (seated) Stephen Dodd; (far right) Ramona Ferenz. (Photo courtesy of MITRE Corporation.)

produced a core memory system for the Mark IV called a magnetic-core delay line (think mercury delay line without the mercury). Putting digital data into a small donut of magnetic material instead of into a huge glass bottle like an electrostatic storage tube was a technological breakthrough with enormous implications. It was the answer to a dependable, cheap, and easy-to-fabricate memory system; it was what enabled the stored-program computer to become a commercial reality.[26]

Wang's problem was that he set out his cores in a series, like a string of Christmas lights, which ultimately proved to be its downfall. Like the antiquated mercury delay line before it that slowly bubbled up data one slow piece at a time, so too did the magnetic-core delay line. For real-time needs like those of Whirlwind, a magnetic-core delay line would never do. On September 29, 1949, just before he patented his core memory invention, Wang presented it as a paper to a Harvard computing symposium. In the audience were bright boys Jay Forrester and William Papian. Although Wang was a PhD with some digital circuit experience, he was not an experienced computer builder like Forrester and Papian. Forrester instantly saw the wrong direction Wang had taken.

Forrester, as well as anyone else who was experienced in building computers, knew immediately that arranging the cores in a series was a backwards step. Accessing data would be stable and safe, but would also be way too slow. Fast was the correct direction for the future. Forrester also determined after testing Deltamax that it was not the right magnetic material to use. He had already bent, kneaded, and shaped Deltamax into rings; he had already experimented running an electric current through the rings; and he had already concluded that the material was too slow at switching current and its physical composition was too sensitive to endure the rigors of mass production.

Forrester laid his cores out in a matrix of 32 cores by 32 cores, and joined them together with interconnecting wires. The matrix design made any core instantly accessible. And it was very fast. The superiority of Forrester's matrix was undeniably self-evident to anyone who saw it. When Wang learned what Forrester had done, he called it "brilliant" and always regretted not thinking of it himself.[27] An Wang took his core magnet patent and started a company in 1951. Forrester didn't. Although, with the walls closing in on the Barta Building, it would have been a perfect opportunity to dash off with a deep-pocketed investor and a couple of sharp lawyers, and hang a sign out that read something like "Jay's Magical Magnets." It would have been hand-over-fist millions for the young engineer, but he stayed put. He thought only about his mates and his machine. With his design set and his proof of concept complete, Forrester next sought out better cores. In 1949 he got Ernst Albers-Schönberg of General Ceramics to brew up hardier magnetic materials. A year later Albers-Schonberg came up with a magnesium-manganese ferrite material that suited all of Forrester's needs.

Curious George

George Valley was on the telephone with Washington for what must have seemed like hours. On the other end were a very interested Theodore von Karman and his aide Air Force Major Ted Walkowicz. It was just about the first of November 1949, as Valley recalled it, and he had just finished with his analysis of the country's air defense preparedness. It wasn't good, and he had an earful of suggestions jumping over the telephone line.[28]

He had been to a Continental Air Command (CONAC) radar site, and he had read all the air defense briefings courtesy of von Karman's Scientific Advisory Board. (Valley had been an active member of the SAB's Electronics Panel beginning in 1946.) He had been to Marchetti's AFCRL where he was briefed again and then shown experimental radars at the AFCRL's airfield in Bedford, Massachusetts. He had seen smatterings of potential amid a

frighteningly nonexistent air defense capability that only served to reinforce Bob
Robertson's scathing assessment. Valley
saw to the heart of the problem and also
saw a solution. He believed that the major
technical problem to be solved was to assess
the "positions of aircraft quickly enough to
predict their future positions."[29] Valley felt
sure that the problem was soluble through
electronics. Valley's observations definitely
got their attention; they "insisted I write
them a letter explaining my thoughts about
air defense in detail."[30]

Close-up of magnetic core memory.
(Photo courtesy of MITRE Corporation.)

Valley's talk of disarray with air defense was not a surprise to von Karman
and Walkowicz. They lived with it daily. Disarray was everywhere in the
newest service, almost since its inception in September of 1947. Not only
were there virtually no air defenses but also the Strategic Air Command
(SAC), the Air Force's supposedly nonpareil offensive capability, was also
woeful. "SAC crews were understaffed and ill prepared for combat missions, bombers did not possess the range to attack the Soviet Union . . .
plans to attack key Soviet military and industrial sites were sketchy at best
because of inadequate intelligence."[31]

When General Curtis LeMay took command in October of 1948, he said
that SAC crews not only could not find Moscow; they could not find St.
Louis. If and when the crews did locate Moscow, they had next to no atomic
bombs at their disposal, which completely shocked President Truman.[32]
Compounding the situation were dispatches from overseas commands regarding highly suspicious behavior on the part of the Soviets. General Ennis
Whitehead, Air Force Commander in the East, noting "strange incidents
and excursions" over Japan, thought for sure that the USSR was getting
ready for war "within a few months," as did General Lucius Clay, Military
Governor of Germany, saying that war "could come with dramatic sud-
denness."[33] World events didn't help either. The Nationalist Chinese were
on the run; Cochin China (Southeast Asia) was exploding; Germany was
experiencing the Berlin Air Lift; Czechoslovakia a Communist coup; and
Greece a civil war. The aftershocks of it all were unnerving the Pentagon
brain trust.

The new Office of the Secretary of Defense (established in 1947 with
James Forrestal as its first Secretary) was also experiencing aftershocks and
trying to anticipate the repercussions. Either because of actual intelligence

information that Soviet clandestine attacks on the United States might be looming, or just to be sure that the potential for such types of attacks was being addressed by America's defense establishment, or because everyone was on edge or paranoid or both or all of the above, Forrestal put together an ultra secret organization to watch out for terrorist assaults on an unsuspecting America.

Back after the war with his Manhattan law firm, Major General William (Wild Bill) Donovan, wartime leader of the Office of Strategic Services (OSS), the precursor to the Central Intelligence Agency, was re-recruited back into government service by Forrestal to lead the ultra secret watchdog group. Donovan and Forrestal assembled the necessary people and dealt out their respective assignments. Vannevar Bush was a member, as were Karl Compton, Admiral Sidney Souers (future Director of the CIA), and General Alfred Gruenther (Director of the Joint Chiefs of Staff). Donovan further created subgroups within the secret organization and staffed them as well. The overall title for the group and its inquiries was the Committee for the Defense Against Unconventional Attacks.[34] By "unconventional" these guys meant to cover just about any kind of terrorist operation, ranging from clandestine attacks using explosives or biological weapons or chemicals or radiological weapons—to economic or psychological attacks or violence against key individuals or groups. Hopefully they were better prepared than air defense or SAC, but that was questionable. Forrestal's own sanity was questionable. He was said to be mentally unraveling, and it got worse as 1948 rolled on.

On March 11, 1948, Forrestal held a military conference in Key West, Florida to iron out the "roles and missions" of the respective services. In other words, he wanted to make clear "who does what with what weapons." The Air Force got the job of continental air defense, including Army anti-aircraft batteries, which miffed the Army; Navy admirals got the go-ahead on their Truman-approved, $139 million supercarrier the USS *United States*, while commitment to build B-36 long-range, strategic bombers for the Air Force was scrapped, which miffed SAC. Either because of the warnings from Japan and Europe, or because of Air Force intelligence thinking that an attack was imminent, or because the Air Force was ruffled over the B-36 outcome at the Key West Conference, General Spaatz on March 25th ordered emergency air defense measures. America's pitiful air defenses went on alert with the guarding of the Hanford Atomic plant in the State of Washington, which was just about the best they could muster. Radars in use were vintage Word War II units—familiar to the Soviets from Lend-Lease equipment during the war—which provided coverage out to 150 miles at 20,000 feet and zero coverage above 35,000 feet.[35] Congress got worried.

In April, in a supposedly tight-fisted economy, the House Appropriations Committee voted an $822 million supplemental appropriation to build out the Air Force's strategic capability, a sum almost equal to six Navy super-carriers. Initially rejected by the Bureau of the Budget and then tabled by the Army and Navy for further review, Air Force Secretary Symington (a former Senator) exercised his considerable political wiles with Congress and got the plan before the House. The budget was introduced into the Senate on May 27th, and into the House on June 2nd. On June 3rd, Truman locked the checkbook and told Forrestal to look for a cheaper alternative.[53] On April 30, 1948, Spaatz retired to the Civil Air Patrol and to military affairs editor of *Newsweek* magazine, and was replaced as Air Force Chief by General Hoyt Vandenberg. The wartime Vandenberg was the organizer of the Twelfth Air Force, and a Silver Star, Distinguished Flying Cross bomber hero flying in the North African campaign. Following the war, he was the first director of the Central Intelligence Group (later renamed the Central Intelligence Agency). The *Washington Post* called him the "most impossibly handsome man in Washington," and Marilyn Monroe called him the man with whom she would love to be marooned on a desert island.

The Air Force next conducted an exercise in the Pacific Northwest from April 28th to May 10, 1948, which displayed further the disaster that was air defense. The F-80 Shooting Star jet fighters showed their lack of range and meager electronics in failing to intercept an attacking force of B-29s. Twin engine, prop-driven P-61s were totally ineffective when unable to close within firing range of the bombers. If they had been Soviet B-29s (TU-4s), Washington and Oregon would have been hard hit.[36] At a second "roles and missions" conference in Newport, Rhode Island that August, the Air Force won operational control of the atomic bomb, something it had long sought.[37] And a bomb, of course, needs a good, long-range bomber to carry it. And such a bomber would be soon in coming after Vandenberg set the table by getting SAC chief General George Kenney to retire, replacing him in October with LeMay. The previous May, Vandenberg brought on his good friend General Muir Fairchild as his vice chief and right-hand man. Called "Santy" by his friends, Fairchild had been an air defense advocate as far back as the late 1930s[38] when he, Vandenberg, Chennault, and Saville were all instructors at the Air Corps Tactical School at Maxwell Field. He wisely subsumed much of that advocacy after witnessing what happened to Chennault and Saville for being too forceful about pushing air defense

General Muir Fairchild. (Photo courtesy of USAF Historical Records Department.)

into the faces of the bomber mafia. Thereafter, he was a bombing protagonist, keeping defense well under wraps until the appropriate time. During the war he actually gained a reputation as well as a few medals for being a brilliant bombing strategist. He probably would never have made it to addressing the SAB that day had he not low keyed defense. Unschooled—he was not a college graduate—Fairchild was considered brilliant, analytical, introspective, the "philosopher" of air power, and a founding father of the Air University.[39] Fairchild, in turn, called upon his old buddy and air-defense genius General Gordon Saville, who had been shunted off to Rio de Janeiro the whole while as liaison to the Brazilian Air Force. When Saville heard that Fairchild had become vice chief, he exclaimed in relief: "Thank God, Santy's there." Saville returned to Washington in June, and on September 9, 1948, he briefed Forrestal, Air Force Secretary Symington, and Air Force staff on just how shockingly bad was continental air defense.[40]

Somewhere in that mix of air defense and ultra secret committees, the utility of a real-time computer like Whirlwind might well have provoked a few thoughts here and there. With MIT guys like Bush and Compton in the inner of inner circles, especially Compton who supposedly wanted to bring L-1, L-2, and L-3 to Washington for the avowed purpose of influencing the influential, 1949 was a perfect opportunity. Or, at the very least, Compton might have phoned Killian (then president of MIT) to say something like, "Hey, I can't be specific and give you an exact reason right yet, but please do whatever it takes to keep Whirlwind healthy. We may have a need for it." None of that happened. No light bulbs were lighting up in their heads. Even with a bunch of military uses for the computer in his hands, Compton never made the connection to push MIT's technological advance in computing into the light of day. Even a meager shout of support from the sidelines would have helped when the Ad Hoc Panel on Electronic Digital Computers started its grand jury-like deliberations on which government computers to axe. The Ad Hoc Panel heard testimony from experts, made investigations, and conducted tours of computer facilities—including a look-see at the bright boys' electronic leviathan—[41]culminating with a fact-finding mission to Harvard's Second Symposium on Large-Scale Digital Calculating Machinery in September of 1949. No experts or character witnesses presented on behalf of Whirlwind. The bright boys must have felt like they had been hung out to dangle in the wind.

In March of 1949, Truman, distrustful of Forrestal's ability to perform, asked for his resignation and replaced him with Louis Johnson, who promptly nixed the supercarrier in favor of building B-36s, which infuriated the admirals. It was pre-Korean War, when the US had few forward bases abroad other than England, Germany, Japan, and the Philippines; [42] long-range, strategic B-36s seemed like a safe alternative. Distraught and apparently demented, a deposed Forrestal was put under psychiatric care, but in May of 1949 leapt to his death from the 16th floor of the Bethesda Naval Hospital. Yes, von Karman and Walkowicz were well acquainted with Valley's observations. Valley, however, just maybe had a potential solution to part of the problem, which came as a bit of fresh air into the stifling atmosphere of an America floundering at preparedness. Von Karman and Walkowicz eagerly awaited Valley's written comments.

They didn't have to wait long. Valley quickly sent along a three-page summary on November 8th. Since his backseat conversation with Bob Robertson, Valley had been quick about whatever he was doing on behalf of air defense, although he was less sure about his motives for jumping in so readily—especially when it came to anything having to do with an atomic bomb. Valley had refused to work on the Manhattan Project and following the war had joined up with others "to make nuclear energy into a force for peace, not doom."[43] He lobbied Congress against the May-Johnson Bill that would have placed nuclear energy entirely under the control of the Department of Defense. He made "innumerable speeches to lawyer's clubs, to doctor's clubs, to chambers of commerce, to Rotary Clubs, to Lions Clubs, to the League of Women Voters, to anybody who would listen."[44] In the end, however, his thoughts were on home and family, and the sobering realization that "the blast wave of the first bomb to hit Boston" would easily reach his almost-completed home in nearby Belmont. So, with more givings than misgivings, he sent his list off to von Karman.

His letter noted ten subject areas to be investigated, and stated that the investigations would best be accomplished not by the military but by a special "civilian" committee drawn up of individuals with expertise in the areas he outlined. For speed and to facilitate the meetings, he suggested that committee members live in close proximity to one another—the New York to Boston corridor—and that the meetings take place in the

Boston–Cambridge area, preferably at the AFCRL.[45] Twenty days later, on November 28th, Valley was in Washington to attend a special meeting of the SAB Executive Committee. Muir Fairchild was there to address the group, and he read excerpts from Valley's letter to von Karman. "I was impressed by his frankness. He asked the SAB to fix the system."[46] It was fortuitous that Valley's recommendations got into the hands of someone like Fairchild. Von Karman had placed Valley's letter with a doer. The next day, the SAB formalized one of Valley's suggestions by proposing that an Air Defense Technical Committee be formed to address the matter. Fairchild made sure that the proposal got quickly to Vandenberg for his review and approval. Things were beginning to look up for air defense.

Things were looking decidedly down for Whirlwind. Two days after the SAB meeting on December 1st, bad news hit the Barta Building with the release of the Ad Hoc Panel's "Report on Electronic Digital Computers by the Consultants to the Chairman of the Research and Development Board." According to the report's preliminary findings and recommendations, Whirlwind and the effort spent building it were "out of all proportion to the effort being expended on other projects having better specified objectives."[47] The report's recommendation: cease funding the machine. In a last-ditch effort to sway the Panel away from its fatal decision, Forrester would hit the typewriter keys once more to produce report L-24, titled "Statement of the Status of Project Whirlwind Prepared for the Research and Development Board."[48] It didn't cut any ice. The Panel's final report and recommendation for the bright boys' machine in June of 1950 would remain unchanged.

If December of 1949 was a bad month for Forrester, it was worse for the Nationalist Chinese, who completed their evacuation to Taiwan. The flag of the People's Republic flew over Beijing. From Korea to East Germany, the year 1949 saw half the world fall under the sway of Communist domination—and it kept on growing. The Pentagon was jittery. So too was J. Edgar Hoover's FBI, which reported that there were Commies among us, in Hollywood no less, as Danny Kaye, Fredric March, John Garfield, Paul Muni, and Edward G. Robinson were said to be card-carrying members of the Communist Party. Several hundred others were uncovered in Truman's own Administration. The Red Scare was beginning in America. The Republican Senator from Wisconsin, Joe McCarthy, was in the wings warming up for his great inquisition—the purges, blacklistings, and humiliations of his infamous Congressional inquiries, popularly known as the McCarthy Hearings, which probed the country for Soviet spies or Communist sympathizers anywhere and everywhere. The 1940 Smith Act made advocacy of seditious ideas a federal crime, criminalized Communist teachings, and became the mandate and cudgel for witch hunts. For the few people whom McCarthy exposed, the reputations, careers, and in some cases the lives of many otherwise innocent citizens were ruined amid the pounding of his gavel.

By 1949 Gordon Saville was back by a year from Brazil, back at Mitchel Field, Long Island, back to where in the 1920s he had first started out as an air cadet and then as a rookie officer. It was a familiar and friendly place for him. He was back where his first boss Lt. Colonel Benjamin Foulois (later Chief of the Air Corps) recognized his exceptional abilities and recommended him for ACTS, where he went on to graduate first in his class. He was back again at Mitchel Field where Foulois and his wife helped Saville court a girl from Connecticut whom he would marry in 1928. Everywhere at Mitchel Field there were hidden memories that now popped back into view. It was also the place where during the war he would first put together his ideas on air defense and interceptor aircraft, ideas that he and Claire Chennault had conjured up in the 1930s at Maxwell Field. The brash Georgia boy, who graduated from and then taught at the Air Corps Tactical School, who later commanded fighters in North Africa, was back to the skies where it all began. His beloved Air Force had changed so much since those early days with Foulois; the Air Force had gone from biplanes to supersonic jets, and then started in on rockets and missiles, and even entertained thoughts about satellites. His buddy Santy Fairchild had reeled him in from Rio in 1948 to work his air defense stratagems for the country. By 1949 Gordon Saville was showing himself as America's premier authority on air defense and would be behind almost every air defense program until his retirement in 1951.[49]

Saville's work began by examining someone else's plan for air defense. It was an ambitiously massive, two-part plan that was worthy of a Roman Emperor. General George Stratemeyer, leader of the Air Defense Command, had concocted the first part regarding the interception and destruction of enemy bombers. It called for "38 control and warning groups, 34 all-weather fighter groups, about 300 antiaircraft artillery battalions, and 83 guided-missile groups, requiring 700,000 people for implementation."[50] Before any of this destructive intercept could be unleashed, enemy bombers would first have to be spotted in what Stratemeyer termed a radar "trip wire," which was the second part of the plan. The "trip wire" part was left to Maj. General Francis Ankenbrandt to prepare. Ankenbrandt felt sure that the best plan to present was one that would "constitute tangible proof to the nation that the Air Force was serious about defending the United States against air attack."[51] Ankenbrandt's plan was indeed serious and huge and expensive. Called the Radar Fence Plan, codenamed Project Supremacy, it called for "411 radar stations and 18 control centers in the continental United States serviced around the clock by 25,000 regular U.S. Air Force personnel and nearly 14,000 Air National Guard radar specialists."[52] Time to build it all: five years; cost: $600 million. Quite the air defense system, especially when the Truman Administration was howling about only minimal military spending. Much of the plan, admitted Ankenbrandt, expected to utilize World War II-era equipment that was deemed obsolete, or declassified, or very well known to the USSR. Hadrian's Wall would never come to be; Truman just said no.[53]

Saville had already anticipated the negative outcome for Project Supremacy. The sheer magnitude of the plan seemed over the top, while structurally it rested on vulnerable, obsolete technology. Unorthodox in everything he ever did, hard driving, having no patience for fools, adept at avoiding red tape, feisty, a skillful leader with over a decade of air defense experience, he dashed toward his concept of a plan that was at once solid, doable, and a bargain. Assigned to Stratemeyer's Air Defense Command, he paid little attention to his boss, except for grabbing off Stratemeyer's best and brightest officers for his own planning needs. Set up as Special Projects Officer and reporting directly and only to Vandenberg and Fairchild, Saville had unusual freedom without any need to heed or follow the traditional chain of command way of doing things. Many the bruised ego and ruffled epaulet followed in his wake.

"I wasn't going to stand in line and wait," he recounted later. "Time was pressing here."[54] He quickly assembled a staff of air defense planners at both Mitchel Field and at the Pentagon, who first made a complete survey of the entire postwar air defense situation. He reasoned that any truly effective plan needed to begin by accounting for radar and its proper positioning, and then everything else would follow. That thinking would hold sway for over a half century. Radar, part of what today General John Piotrowski, former Commander in Chief of Air Force Space Command calls "sensor systems," comprises first contact with the enemy and as such becomes "the first line of defense" in what he terms truly effective "synchronized offensive and defensive forces."[55] In two months Saville's plan was complete. He recommended "75 radar stations and 10 control centers in the continental United States, with 10 radar stations and 1 control center in Alaska."[56] At $116 million, it was a sixth of Project Supremacy's price tag, and it could be built over a two-year period. There were even ideas about research dollars for new technology and advanced radars. As Vandenberg wrote later in his journal, Saville exhorted Fairchild and me to take quick action, saying

"Nothing can be found in the world situation, in the attitude of the people, or in any other field which would justify continued delay. We must get on with it."[57] The first phase would be the immediate installation of a very temporary radar system that he called the "Lash Up" system (each radar was literally "lashed" atop of a tower or platform). Then, as quickly as possible, the permanent system would then begin to replace all of the lashed-up radars. The Air Force then combined the Tactical Air Command (TAC) and the Air Defense Command (ADC) into a single entity called Continental Air Command (CONAC). To placate a particularly bruised ego, Stratemeyer was given command of CONAC, with the subordinate command of ADC going to Saville. Fairchild and Saville next took the proposal to Congress. In rapid succession Congress approved the plan and Truman signed it on March 21, 1949.[58]

Fairchild and Saville were not finished. With a plan to replace Ankenbrandt's "air defense" done, funded, and ready to implement, they next took on how best to attack enemy bomber formations. Stratemeyer's plan for "intercept and destruction" needed a fresh look. As Saville had done during the war when he invented the interceptor category and got the P-38 produced to fill that need, the fighter pilot in him went looking again for a suitable interceptor. He recruited two old air buddies to help him out: the experienced fighter pilot, Col. Bruce Holloway and the veteran fighter commander, General William Kepner, head of the Air Proving Ground at Eglin Air Force Base in Florida. These three Air Force bright boys, together with Fairchild, would set the direction for modern fighter aircraft for the next decade. They regarded the existing F-82 Twin Mustang and the F-61 as dismal, and they referred to the jet fighters, Republic's F-84 Thunderjet and Lockheed's F-80 Shooting Star, as "practically worthless."[59] Dismissing the conventional wisdom of retaining these aircraft for the foreseeable future, they cut in an entirely new direction. They sought out another like-minded maverick in James H. "Dutch" Kindelberger of North American Aviation (inventor of the P-51 Mustang), who urged them to transform his F-86 Sabre jet into an all-weather night fighter and interceptor. They wanted the F-86 to "take off from a runway when the ceiling was zero-zero (no visibility), go and make the interception and then get back on the runway." They further asked that the pilot operate both the flight controls as well as the radar (other aircraft of the time had both a pilot and a radar operator). Kindelberger took on the conversion, and it proved so successful that the Air Force ordered 124 of the new F-86D Sabre jets as air defense interceptors.[60]

For the future, they set into motion a design competition for a new jet fighter to be ready by 1954, which they called the "1954 interceptor." That aircraft turned out to be the F-102 Delta Dagger, the country's first delta-winged aircraft, and the first of the famous Century Series of advanced performance and avionics fighters that were continuously refined, redesigned, and built well into the 1960s. Fairchild further did away with rigid specifications for fighters, as had been standard procedure with all former Air Force design requests. He had Saville brief manufacturers with the "defense situation and outline general requirements" only; the industry was then free to interpret the task at hand. He insisted also that the new design be built from a "weapons system" point of view—which is still in effect today—treating the aircraft as a single, integrated entity and not as a bunch of individual components.

By the spring of 1949 Muir Fairchild and Gordon Saville had $116 million to spend on continental air defense; at the same moment in Cambridge, Jay Forrester was pondering Deltamax, jotting down matrix patterns in his notebook, and keeping a watchful, anxious eye on the calendar. The air defense plan, however, had many flaws of which the Air Force was well aware. A Ground Observer Corps stationed in towers with binoculars to spot planes under 5,000 feet would be ineffective, and Saville knew it all too well from his World War II days. Then too, humans were overmatched using grease pencils to scribble on Plexiglas walls the positioning information for waves of incoming planes. Bob Robertson noticed the deficits in the system; he railed about them to Valley during the taxi ride.

Hastily thrown together radar sites were what Valley analyzed on his quick tour, and what he elaborated on in his letter to von Karman. The Air Force understood that engineers and physicists needed to have at the plan to make it work; they too had learned well from their OSRD and Rad Lab wartime experiences. Guy Stever points out that "the enormous success of wartime research through the OSRD model made the military and the American public passionate believers in high-quality science."[61] With a rudimentary air defense system in place backed by adequate funding, the Air Force could then afford itself the luxury of tinkering the system through successive upgrades.

In 1949 there was no OSRD or Rad Lab to do the work, and the Research and Development Board was not set up to be an OSRD substitute. "The three services fervently believed that if the OSRD went, something had to rise in its place."[62] The Navy jumped in first in 1946, forming the Office of Naval Research (ONR). Hap Arnold and von Karman bet on the SAB to be their OSRD replacement. Valley, as a SAB member working proactively

George Valley, Director of SAGE, 1952–1958, speaking to members of the press at announcement of SAGE system for continental air defense. (Photo courtesy of MITRE Corporation.)

on his air defense analysis for the Air Force, fulfilled Arnold's faith in the organization. That Valley acted on his own personal initiative and not at the SAB's request would work wonders at building and maintaining trust between scientists and airmen. Valley's timing was good. The Department of Defense anointed the Air Force with the task of continental air defense on March 11, 1948; Saville briefed Forrestal and Secretary Symington on September 9th; and by March of the following year the Air Force had a plan in hand and money to start the job. Five months on and the Soviet A-bomb went off. Truman revealed it to the nation in September; and on November 27, 1949, Muir Fairchild was before the SAB asking for its scientific and engineering help.

By December 8th, Fairchild had forwarded the SAB-Valley proposal for an Air Defense Committee through to Vandenberg.[63] A letter from Fairchild to Valley on December 15th confirmed that the Committee got a strong go-ahead from the Chief as well as a request to begin work "within the next few weeks."[64] Another letter from Fairchild of the same date asked Valley to take on the chairmanship of the Air Defense Committee. The letter also included the names of six others who would be active Committee members; among the names was Valley's chess partner from their rides together on the London–Great Malvern route, Guy Stever. That same day, Valley wrote a four-page paper elaborating on his original letter to von Karman. "Tentative Remarks on the Task, Organization, and Program of the SAB" reiterated the ten subject areas to be investigated, and with it the SAB formed the Air Defense System Engineering Committee or, as it was commonly called, ADSEC. Similar to A. P. Rowe's Sunday Soviets of mixed military officers, engineers, and scientists meeting at the Grosvenor Hotel during the war, ADSEC had a similar makeup of guests but met on Fridays (actually the third Friday of every month) in decidedly less posh surroundings at the AFCRL. However, the intent was identical: to put a diverse group of heads together bent on a common goal—air defense.

The Gang's All Here

Serendipity stalked George Valley during December of 1949. Although his thoughts were squarely on radars, there was an ever-enlarging empty spot of concern in his radar plans; a baffling problem that sat in the middle of everything. The air defense dilemma facing Valley and Marchetti had two parts. First, how to efficiently and very quickly transport a glut of radar information from one or multiple radar sites to a control center. And secondly, once the information got to the control center, how then to reassemble the information into a comprehensive picture of an overall air situation. Whether for Ankenbrandt's 411 radars or for Saville's 75 radars, or even for a single radar, the old way, by telephone, was impossibly slow and too laborious to grease pencil telephoned coordinates onto a Plexiglas wall. Utter bedlam would be the result of radar operators shouting coordinates over telephones for wave upon wave of incoming enemy bombers.

Also gnawing at the pair was the vulnerability to low-level attacks coming under the radar screen. Long-range surveillance radars, like the AN/CPS-5 (generally referred to as Microwave Early Warning or MEW radar) had an effective range of 60 miles up to 35,000 feet. In good weather they might scout 200 miles out. However, in order to avoid clutter and false returns from buildings, mountains, and other terrestrial objects, these radars were tilted upwards. As such, 5,000 feet of air space was left exposed to low-level air attack. The solution of volunteers by the thousands manning Ground Observer Corps towers relaying information from binocular sightings, especially at night, was too frightening to think about. Also, with the near-term likelihood of jet aircraft replacing prop bombers, 60 or even 200 miles of warning would be insufficient. Beginning in December of 1949, 24 locations on Saville's list of 75 were turned into temporary Lash Up sites,[65] each with a CPS-5 (distance finding) as well as a TPS-10 height finder (altitude finding) radar. ADSEC's job would be to come up with a plan to convert all 75 locations into a permanent air defense perimeter; permanent . . . and certainly more capable than the Lash Up sites. Valley and Marchetti were at an impasse as to exactly how to pull that off.

Part of the answer already existed at Marchetti's laboratory. One of his AFCRL bright boys, John Harrington, had been tinkering with equipment that might help. Harrington's Relay Systems Laboratory in 1948 had relayed by microwave a radar video picture 20 miles from Hanscom Field to the AFCRL and reconstructed it on a display screen called a plan-position indicator or PPI.[66] Unlike today, microwave towers in the late 1940s were a distinct rarity, expensive to build and to maintain. Harrington was urged to find another way to relay the radar sightings. Telephone lines were suggested because they "were abundant, available everywhere within the United States, and much less expensive than microwave transmission."[67] The problem was that thin copper telephone lines carried little information compared to microwaves traveling tower to tower through the air. Harrington theorized that by utilizing the new technology of binary digital transmission, wideband microwave could be replaced with narrowband telephone lines. He would need to convert analog radar data into 1s and 0s, send it over noisy telephone lines, and then reconvert the data back into an analog signal for a CRT (cathode ray tube) display at the AFCRL. His efforts paid off with the development of digital radar relay or DRR, which in late 1949 he tested by sending radar information from MEW radar at Hanscom to a display scope in the AFCRL.[68] DRR still needed work, but Harrington and his fellow bright boys had pulled off a world's first in automatically detecting radar targets, encoding target coordinates, storing them, and then transmitting them—albeit at a low rate—over telephone lines to a remote site for display or for further processing.[69]

Having seen Harrington's DRR system at work, Valley, one evening late in the same year, hit upon a solution for low-level, under 5,000-foot detection. Doodling while correcting student papers, he scribbled out a triangulation pattern for three Continuous Wave (CW) radars to detect the position of low-flying aircraft. CW radars, as their name implies, continuously emit a microwave beam. They were small, like the size of the magnetron wave generator in a microwave oven. "I quickly discovered that three CW radars, emplaced at different known points [atop telephone poles], could provide signals from which the position of a target seen by all of them could be computed."[70] By "compute," Valley was thinking of a human operator leaning over a display screen with a book of logarithms and slide rule in hand. Marchetti loved the idea, but both of the men knew that the amount of radar data continuously bombarding a human being from CW radars would be overwhelming. "Well, I added lamely," recalled Valley, "maybe we could feed the output of the scaling circuits into a digital computer, and . . . "[71] He was cut short by Marchetti, "Now you're talking, George!" The thought of letting a computer do all the counting was like a revelation to Marchetti, although it should not have been so novel an idea since he had already chatted about computer tracking with Forrester months earlier. But, at least they now felt free to CW as many telephone poles as necessary. That was a Saturday in December of 1949, and by noon, they had worked out a diagram with CW radars spaced every ten miles.[72]

They next began the hunt for a digital computer to do all the counting, which would prove to be a very disappointing adventure. "We did some phoning the following week," to companies, universities, and others known to have built a computer, or to be in the process of building one. "The name Whirlwind was not mentioned by any of my informants," and "their replies were uniformly discouraging: too much time and too much palaver, not to mention the astronomical prices they quoted."[73] Strangely, Marchetti didn't think to mention his visit to the nearby Barta Building. In calling around, Valley was struck by the fact that hardly anyone thought about connecting a computer to the real world. Most used their machine to perform mathematical problem solving or to manipulate lists. Further, he found none had addressed the need of converting analog radar signals into digital data.[74] With an ADSEC meeting scheduled for January 20, 1950, Valley and Marchetti wanted to bring to the party a digital computer ready and available to connect to Harrington's DRR system. They resigned themselves to attend that meeting empty-handed.

It would have been a grand opening to Saville's assault on the air defense problem and a wonderful announcement with which to have heralded the upcoming decade of the 1950s. Just the sound of the words, "the 1950s," had a fresh, newly minted ring to it. And 1950 was the decade's first marvelous strut on stage. The vision of that first step seemed somehow to shake off much of the dark patina of gloom and warfare from the previous decade. Anticipation for it seemed to be building throughout 1949. From coast to coast the country seemed primed for something big to happen. The aspect of a new decade prompted a sense of urgency about getting on with life now that the war's end was four years back. Certainly enough time for the war weary to have shaken off the fact that they had survived it all; enough time to have gone the GI Bill route, selected a career or a means of earning a livelihood, found a mate, a nice home in a nice neighborhood for 2.4 people, begun a baby boom, and carved out weekends for coffee over the funny pages and Sunday rides in the family Plymouth.

The United States was like a rocket—engine ignited, straining in its gantry, and ready for liftoff. Europe, on the other hand, was looking at another miserable decade. England had just stopped rationing clothes, bulldozers were still clearing rubble from cities on the Continent, the Red Army was still threatening every border, *The Third Man*'s Harry Lime was skulking the ruins of Vienna selling bad penicillin on the black market, and George Orwell's Winston Smith got a job from Big Brother rearranging history at the Ministry of Truth in *1984*. Poor old Europe, squat in the middle of the killing fields for six years, was devastated, staggered near mortally, and struggling with all its might to coax a thin flame of hope into illuminating an entire continent of woe. Europe was waiting for a miracle to happen, America was making one.

America was Broadway's *South Pacific*, splashing itself over the country with excitement and verve as Mary Martin sang about washing "that man" and everything else right out of her hair. America was agog with the exotic splendor and vivacity of faraway places with funny names from where GIs had just returned. Adapted from James Michener's best-selling, Pulitzer Prize winner, *Tales of the South Pacific*, the play like the novel was set in World War II Polynesia, and adapted to the stage by Rodgers and Hammerstein. Broadway transformed the grisly experience of war in the Pacific into musical theatre of palm trees, warm breezes, luaus, romance, and the crooning of Ezio Pinza. Europe's uncertainty about itself dripped from the eloquent fatalism of Harry Lime's famous lines: "In Italy for thirty years under the Borgias they had warfare, terror, murder, bloodshed—but they produced Michelangelo, Leornardo da Vinci, and the Renaissance. In Switzerland they had brotherly love, five hundred years of democracy and what did that produce—the cuckoo clock." America was a freshly laundered Hawaiian shirt with nothing standing in its way. And America would have over three decades more with which to saunter through it all before the rest of the world caught up. No one figured on the North Koreans coming along to put a few major wrinkles into that crisp shirt.

America's break into the open was led by technology, and its tremendous potential to create wealth. Temptingly, it beckoned to those who recognized the opportunity and seized upon it. Thereafter, fabulous technological innovations begat fabulous fortunes that begat the mega-giants of corporate America. Electronics led the pack, and there was no better place to witness the parade going by than from the sidewalk in front of the Barta Building. Word of the future king, the transistor, was on the street. When announced at Bell Labs' headquarters the previous June, it made hardly a ripple of news, winding up buried way inside the *New York Times*. That the thumbnail size device could substitute for an electron tube, require no hot filament and last nearly forever was a big event for the bright boys. It filled a huge hole in the future of computing by validating that the shakiest part of an electronic digital computer, its banks of unreliable glowing tubes, had a solution that was only a matter of time in coming to pass. First used in hearing aids and transistor radios, the transistor would take a few years before wending its way into a computer. A 1949 MIT graduate student, Robert Noyce, would take considerable notice. Noyce would later gang transistors together to co-invent the integrated circuit, the forerunner of modern microprocessors—the stuff that put the boom into Silicon Valley.

Information theory rumbled down Massachusetts Avenue as Claude Shannon's "A Mathematical Theory of Communication" showed that "something as seemingly abstract as information could be measured and quantified—that it is intimately linked to thermodynamics, that Nature seems to speak in the language of information,"[75] that radar signals, voices over a telephone, and the data coursing through Whirlwind's heart were all the same. When Valley witnessed Harrington's work with the digital transmission of 1s and 0s as radar pictures over voice telephone lines, he remarked that it "was the first practical demonstration of information theory that I had seen."[76] Norbert Wiener's *Cybernetics* showed the relationship between information and control, and that the information in what he called "feedback"—whether in humans, animals, or machines—contained "messages" used for control. His *cybernaut*, the Greek word meaning "steersman," uses the message information from feedback to "steer" "through the random forces of the physical world, based on information about the past and forecasts about the future."[77] The works of both Shannon and Wiener are marvelously brought together by Howard Rheingold in his *Tools for Thought*:

> Common words like *communication* and *message* were given new, technical meanings by Wiener and Claude Shannon, who independently and roughly simultaneously demonstrated that everything from the random motions of subatomic particles to the behavior of electrical switching networks and the intelligibility of human speech is related in a way that can be expressed through certain basic mathematical equations.[78]

What wonderful New Year's presents for the bright boys were these lines of convergence. The boys had grabbed onto the right star and were on the ride of their lives. Their work was as new as new can be and astoundingly important to boot. They were at the heart of something really big, manipulating Shannon and Wiener's work on a daily basis. All the pieces of the digital future were tumbling together around them. These bright boys were defining a new industry and giving themselves new careers within it. As with most new things, not many others could see it as well and many more not at all. The bright boys had the digital eyes. Everyone else was on the same sidewalk but sporting 20/20 vision and seeing only traffic jams and pedestrians.

The New Year put no bounce into George Valley's step as he shuffled through the halls of MIT late in January of 1950. He and Marchetti were still without a computer. It was discouraging. Bumping into old friend Jerome Wiesner in the hallway was just the elixir he needed to get revved up again. Wiesner, at 35 years old, was an electrical engineer in the Research Laboratory of Electronics. He was a Rad Lab grad, an expert in microwaves and radar, a decade distant from becoming science advisor to President Kennedy, and a good couple of decades distant from becoming MIT's 13th president. Valley's run-in with Jerry, as he was called by his buddies, was most fortuitous; he could not have planned it any better. Wiesner was brilliant, compassionate, saw the big picture in things, was no slave to accepted opinion, and loved new ideas and breakaway thinking. George's tale of radars, computers, and air defense got Wiesner's motor running.

"I told him about connecting radars and computers, and that I could get money to make a test if I could find a computer whose proprietors weren't too crazy or too busy. He immediately replied that one was up for grabs, right there on the MIT campus."[79] Wiesner then went on to add the name Forrester to the name of a machine called Whirlwind and that both haunted a place called the Barta Building, which Valley passed often because it "was halfway between my office and Marchetti's." Wiesner didn't tell him much else, nothing about Forrester's problems or the machine's downside. Wiesner did one better; he arranged a lunch where Valley and Forrester could meet face-to-face and figure things out for themselves.[80]

Several days thereafter, on the Friday afternoon of January 27th, Valley and Forrester probed one another over warm dinner rolls, while Jerry Wiesner, his matchmaking completed, faded quietly back to the halls of MIT.

What Valley was up to did not come as a surprise to Forrester. Seven days earlier, Perry Crawford, while visiting Whirlwind, told him that ADSEC had recently gotten underway. ADSEC, in fact, was very quickly taking on a new, unofficial moniker as the Valley Committee. Forrester also had a pretty good idea as to its mission for the Air Force. Valley had done some sleuthing as well. He had checked around about Forrester and his machine: "All these people gave us reports that differed only in their degree of negativity."[81] Marchetti called the ONR to get more information, which hinted that the ONR was ready to drop the machine entirely. Marchetti and Valley were undaunted by the stories circulating about the Barta Building

Interior of Barta Building and Whirlwind memory unit: Charles Corderman (on ladder); Gus O'Brien (below left); and Norm Daggett (below right). (Photo courtesy of MITRE Corporation.)

as a house of horrors. Valley plainly and completely laid out for Forrester all the specifics. Valley's need for Whirlwind must have done wonders for Forrester's blood pressure. What a godsend for a harried heart. It wasn't he and Everett beating every bush from Cambridge to Washington to find a willing sponsor for their machine; it was refreshingly quite the opposite: two men breaking bread together on a wintry Friday afternoon honestly trying to mesh their talent with their hopes in order to find success together. Forrester wrote later in his notebook that Valley seemed truly interested.[82] Such honesty and frankness were also refreshing to him.

Not wanting to spook Valley's interest on a first date, Forrester tiptoed around the ever-present issue of Whirlwind's memory. Everything else about the machine seemed to be winging along except memory. Electrostatic memory tubes had been the machine's bane since Forrester's first investigations with the Panel on Electron Tubes back in 1946. And here he was four years on still fumbling at a solution. He had checked on Raytheon's storage tube project, Edgerton's glow discharge tube, and the famous Williams-Kilburn tube, and by 1949 was making his own for the machine.[83] Of the first 20 tubes made only 13 passed muster. A current cost analysis of tubes with storage capacity of 256 bits each was projected to be $1,500 per tube at a production rate of one and a half tubes per week.[84] Although he was writing in his journal about magnetic-core memory throughout 1949, the most reliable memory around was the antiquated but workable flip-flop memory. Flip-flops were as old as the Eccles-Jordan circuit from 1919. Flip-flop memory was basically two electron tubes that could flip-flop their states from binary 0 to 1 and back again. In terms of storage a flip-flop was a 1-bit memory device. It would take 256 flip-flops to equal one of Forrester's electrostatic tubes. That was a horrid thought.

Worse, if the current was turned off, all memory was lost as well. Flip-flops as switches for Whirlwind were passable devices, but as memory they were worse than a finger in a dike. By March of 1950, the bright boys had 16 of their own hand-built, electrostatic storage tubes in Whirlwind. For a group that prided itself on totally reliable components, these memory tubes were an exasperating reality. Whirlwind as well as every other electronic digital computer had somehow to ditch glass bottles forever. At lunch with George Valley was the big break that Forrester and his bright boys had long sought. He now had to call out the inventor in himself and quickly get about pulling the rabbit of core memory out of his hat.

Forrester invited Valley back to the Barta Building. Valley was overwhelmed by the size of Whirlwind; he had never seen that many electron tubes in his life. And the machine "was functioning: it was calculating a freshman mechanics problem and displaying the solution on a cathode-ray tube."[85] He was hooked. Forrester and Everett loaded him down with the L-1 and L-2 reports, quarterly reports on the Watson Laboratories air traffic control work, memoranda by Gordon Welchman, and David Israel's proposal for his master's thesis on air traffic control.[86] Valley made an appointment to return the following Monday with a few of his friends and associates from ADSEC.

Marchetti wanted to know all about the visit. Was the machine as bad as everyone had told them it was? "I replied that while Whirlwind gave the appearance of being mechanically overdesigned, and also looked like something guaranteed to set the teeth of experienced pulse-circuit designers on edge, it nevertheless seemed to work, and it was available."[87] Valley spent all of Sunday going over the tomes that he lugged out of the Barta Building. He was impressed with the minds that had authored them.

On Monday, January 30th, Valley returned to the Barta Building, accompanied by Marchetti, Eugene Grant, Charles Draper, Guy Stever, and Saville's checkbook. Everett and Forrester got Whirlwind to perform. They all nodded, shook hands—and the air defense of North America began.

Chapter Seven

Island in the Stream

It may not be a very cheerful thought but the Reds right now have about 1,000 bombers that are quite capable of destroying at least 89 American cities in one raid.

—Excerpt from 1951 radio commercial recruiting volunteers for the Ground Observer Corps.[1]

The gang was all there. In a most improbable place a most improbable convergence would come to pass. Who could have reckoned that all the right pieces from all the right people would each find its own way to the drab old building on Massachusetts Avenue at exactly the right time? Who among them could foresee that what each could never do alone, they would now do together, and in the process become a juggernaut of innovation and change?

Marchetti would roll out his radars. His colleague, John Harrington, would then hook them together with his Digital Radar Relay, converting digital signals into analog and then back into digital, sending them over telephone lines from Bedford to Cambridge. Whirlwind could then gorge itself on the digital feast, spitting back aircraft position information onto radarscopes. Saville and Fairchild would keep the Air Force in play and the checkbook open. And Valley, as technology impresario, would manage everything steadily up the totem pole of awareness and acceptance. Their air defense system would blossom into the world's first great electronic system and come to be known as SAGE, for Semi-Automatic Ground Environment. It would be an electronic system like none other before it: a veritable Book of Genesis for all systems builders to come. With Whirlwind the boys wrought the first digital network;[2] the first manifest display of information

Light gun and SAGE tracking scope. (Photo courtesy of MITRE Corporation.)

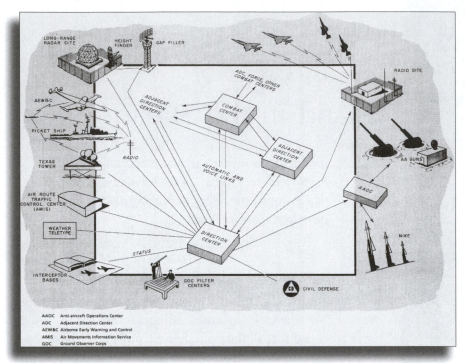

SAGE information network schematic. (Image courtesy of MITRE Corporation.)

theory; and the first coming of the now familiar tools of Information Technology: digital computer, keyboard, light gun (early mouse), monitor screens, printer, software programs, real-time memory, modems, and network connections. With SAGE, the boys would bust out of the Barta Building to build the first information empire. And with it the age of high-tech warfare would begin,[3] and *The Art of War* would take a huge jump up.

When yoked to its giant peripherals—microwave radars from the bright boys at the AFCRL, the network-within-a-building that was Whirlwind would plump out into a network-within-a-region, and later, swell into a network-within-a-continent. By any measure, it was a mind-boggling expansion, especially given the fragile state of digital electronics in the early 1950s. From a single Whirlwind behemoth, 48 advanced replicas of the 55,000- vacuum tube, 32-bit computer were cloned (redesignated in military nomenclature as the AN/FSQ-7). Two each were housed in 23 four-story concrete bastions (plus one more duplex set some 700 feet underground at North Bay, Canada), each 150' by 150' square with 6' thick walls and covering 130,000 feet of floor space, each with its own equally gigantic power and cooling systems, each consuming a white-hot three megawatts of electricity, and each sprinkled strategically all over the country; enough to provide air defense coverage for both the United States and Canada.[4] Linked together by dedicated telephone lines, eerily lit blue rooms of glowing monitor screens tracked everything moving in the sky, fielded calls from 8,000 Ground Observer Corps outposts; and monitored search radars aboard Navy picket ships, on airborne Warning Star radar aircraft, and on oil rig-type ocean platforms called Texas Towers. Across Canada, three continent-wide belts of radar installations at 49, 55, and 70 degrees north latitude—the northern most being the famous 63-station Distant Early Warning (DEW) Line from western Alaska eastward across Canada to Greenland—fed radar data into the SAGE system. In all, over 181 manned and unmanned radar sites, each with either AN/FPS-19 L-band search radar or AN/FPS-124 Doppler radars for low-flying targets, chirped aircraft sightings southward into the giant AN/FSQ-7 computers for processing.

Once intruders were spotted SAGE could also call out interceptor aircraft. The system "was designed to control the interceptor after takeoff, direct it to the target, and bring it back to a final approach to its runway."[5] All of this round-the-clock tracking, reporting, and interception, however, was expensive: the telephone bill alone for SAGE's private, land-leased lines would by 1959 approach $200 million a year.[6] SAGE would be an expensive but versatile sky warrior. It also tracked and reported weather conditions, controlled interceptions using NIKE surface-to-air missiles, and later, targeted 500 BOMARC high-altitude, surface-to-air missiles from ten launch sites, which, at $14 billion for the BOMARCs[7] (built by Boeing), was nearly twice that of the entire SAGE system, at $8 billion.[8]

The bright boys even acted as troubleshooters for Boeing's electronics. One critical discovery early on was that the BOMARC's (IM-99A) test-to-operate switch was kluge, that switching a single rocket to erect and fire actually switched all the rockets to erect and fire. Boeing was quick to fix the situation, recalled bright boy Les Earnest, who discovered the switch problem, wryly recounting how he "published an analysis . . . with the provocative title 'Inadvertent Erection of the IM-99A.'"[9]

The SAGE system, as with the Chain Home filtering system at Bawdsey Manor during World War II, would also present its builders with its own set of unintended consequences. In 1954, General Curtis LeMay, commander of the Strategic Air Command (SAC), recognized the implications of SAGE's real-time information flow for command and control. He opted to dump SAC's slow teletype circuits and radio links.[10] A SAGE descendant, the 465L system, was built for the Air Force and tucked away into the bowels of Colorado's Cheyenne Mountain, home today of NORAD, North American Air Defense. LeMay's enlightenment thrust the Air Force into an upward spiral of ever-improved, real-time information systems, leading to the World Wide Military Command and Control System, the Strategic Automated Command and Control System,[11] and eventually to the early glimmerings of the Internet. In the process the Air Force was "completely transformed from a group that initially resisted computerization to the leading advocate of computerization within the military."[12]

Indeed, there were unseen worlds within Whirlwind. Bob Everett, his crew's carefully crafted report *Whirlwind I Computer Block Diagrams* from 1947, and the bright boys' prudent quality control during Whirlwind's manufacture and management would be poised to explode into a Big Bang of information networking. Peripherals for the bright boys to connect up would number in the thousands, including far distant radar sites, communications systems, and multiple SAGE blockhouses spread out over millions of square miles of geography. Each device, from the smallest modem and light gun to racks of telephone switching bays to huge search radars— every one of which possessed its own electronic subsystem—would all be knitted together into an immense electronic tapestry, a system of systems the likes of which had heretofore never been contemplated much less ever attempted.

Where once there were but a handful of computer programmers in the entire United States, thousands were hired and trained by the System Development Corporation (SDC)—a SAGE subcontractor—to build SAGE's command and control program,[13] prompting SDC's later claim that it alone trained the country's programming industry.[14] SAGE was the first system to innovate computer time-sharing, where multiple computer consoles could simultaneously share the same machine. When SAGE operators used a light gun (like a modern mouse and cursor) to touch a monitor screen and transfer information into or out of computer memory,[15] it was the first time humans directly manipulated a computer; exactly what Licklider envisioned in 1960 with his "man-computer symbiosis." If the bright boys' dream was to build a control computer and not just a number crunching machine, then this was it in spades.

SAGE would not be perfect; it would be flawed and vulnerable and highly controversial, yet there it would be in the flesh for all the world to see, a giant loom spinning information nonstop across a continent and seemingly everywhere in between. This was a system to ape for an as yet unnamed Information Technology; a model upon which an industry could arise. Two years removed from Claude Shannon's famous paper, follow-up work about the true nature of information would just begin to twitch in its chrysalis with the likes of physicist Donald MacKay at the first London Symposium on Information Theory in 1950.[16]

The machine side of real-time information processing, however, would still be a chancy business and lag behind Whirlwind. As late as 1961 real-time digital hardware was still being approached with trepidation, as Sylvania engineer W. A. Hosier cautioned in his apprehensive paper, "Pitfalls and Safeguards in Real-Time Digital Systems." Real-time computers, he wrote, "are among the most complex and delicately coordinated of modern engineering enterprises and involve a wide variety of equipment and techniques that frequently tax the state of their respective arts."[17] He reasoned that if real time wasn't necessary for the computational tasks at hand, then why bother? Most didn't. Whirlwind got lucky; it had found SAGE and found in the military a need for fast information. The world would take some time catching up. However, SAGE would sooner than later spawn totally new industries, open new worlds for scads of manufacturers, and hurtle thousands of individuals into totally new careers. The bright boys created an island in the stream of electronics; an island that would in time become the land.

Civilian authorities similarly recognized an important ability in SAGE: if SAGE could spot every "unfriendly" aircraft over North America, it must then be able to spot and track every "friendly" aircraft as well. Tracking commercial airliners, it was reasoned, might keep them apart and thus avoid midair collisions.[18] *Time* magazine reported in early 1957 that "in the last four months of 1956 . . . there were 452 near-misses between airborne planes."[19] It was only a matter of time until a near-miss became a catastrophe. That same year, a midair collision over the Grand Canyon between United Airlines Flight 718 and Trans World Airlines Flight 2 killed all 128 aboard both aircraft. Enter SAGE, opening the skies to a new way of controlling and managing commercial air traffic, from which modern air traffic control directly descends. The Federal Aviation Administration (FAA) would eventually build its own SAGE-like system to control the national airspace, which, in turn, also became the model for international air traffic control. A SAGE control center in Montana shared its computer and display screens with the FAA for nearly 15 years, serving as the FAA's Great Falls Air Route Traffic Control Center.

IBM, manufacturer of the SAGE AN/FSQ-7 computers, not only pulled in $500 million for building the computers (its biggest single contract of the

1950s),[20] but also built the first electronic digital computer assembly line at IBM's Kingston, New York plant, and an electronic development laboratory in Poughkeepsie, New York.[21] Between 1955 and 1957, IBM employed over 7,000 workers to build SAGE machines. IBM also used its SAGE real-time expertise to push out the world's first airline flight-booking system, the Semi-Automated Business Research Environment (SABRE) system, in which Forrester's old pal Perry Crawford was ensconced helping out the IBM effort. Airline reservations that previously took nearly three hours each to book were made in real time as the SABRE system prowled its database to match flights with customers. The invention of the SAGE database, as well as the coining of the word "database" itself, were two other firsts claimed by the System Development Corporation. "SAGE had to present an up-to-date and consistent representation of the various bombers, fighters and bases to all its users. The System Development Corporation, a RAND Corporation group spun-off to develop the software for SAGE, had adopted the term 'database' to describe the shared collection of data on which all these views were based."[22] Today, of course, every information system has a database, and databases are central to the running of most any business. SAGE not only qualified SDC programmers but also trained hundreds from Western Electric at Murphy Army Hospital in Waltham, Massachusetts to become part of 50-member test teams that went from SAGE blockhouse to blockhouse getting the sites' electronics up and running.[23]

The bottom line to the outpouring of "SAGE effects" upon the new electronics industry seems to have been this: create an exceptional tool and exceptional people will flock to it and use it to create still other exceptional tools. SAGE was just such a tool. The list of supplicants to the SAGE shrine that became digital converts included not only IBM, SDC, and Western Electric and its parent AT&T, but also Burroughs (later merging with Sperry Corporation to become Unisys), General Electric, Raytheon, Philco (later bought by Ford Motor to become Ford Aerospace), Sylvania, and Bendix (later acquired by Control Data Corporation).[24]

Les Earnest reckons that SAGE "was about 10 years ahead of general purpose timesharing systems and 20 years ahead of personal computers and workstations."[25] In the reckless pace that is technological change, that's some kind of longevity. All of which begs the obvious question about why the Navy, the Office of Naval Research, Mina Rees, and MIT were so myopic that they acted to stop the bright boys cold? Their actions were especially strange since they were not callous corporate stockholders hell-bent only on real-time, tangible profits and as such quick to vote out a risky loser. Rather, this group—consisting of the Navy, the Navy's congressionally anointed research laboratory, a vaunted engineering university, and a mathematician tasked with developing computer technology—was supposed to be flexible, open to change, and future looking. In March of 1950, when needed the most, they showed themselves to be otherwise. Absent George Valley and the United States Air Force, who could have guessed where gumption alone would have carried the bright boys? Years spent breathing life into a new technology, a dream project that comes along very infrequently in researchers' careers, would most likely have been dashed forever, no matter how bright the bright boys were.

Decades later in a speech at Boston's Computer Museum, former bright boy Bob Wieser recollected SAGE as a once-in-a-lifetime experience for any engineer. "We had the engineer's dream," he recalled, "a nationally important problem that was interesting and difficult but not impossible to solve. We were in a day-to-day contest with Mother Nature. The odds were bad, but we always had a chance to win, and we won all the battles that led up to SAGE."[26] Wieser even chimed in with his bright boy assessment of Whirlwind's effect upon the then nascent electronics industry: "We also won the cause for digital computing," he reminded his audience. "If there's anyone who thinks we didn't win, just go to Radio Shack and try to buy an analog computer."[27]

Popular culture got an eyeful of SAGE when Hollywood went ga-ga over its mammoth hardware, bulging cabinetry, and flashing lights. No fewer than 34 Hollywood movies and TV programs utilized chunks of SAGE as backdrops and props.[28] What was in the public's imagination as to what a computer should look like and how it should act was totally SAGE transplanted to the sets of *Voyage to the Bottom of the Sea, Lost in Space, The Time Tunnel, The Man from U.N.C.L.E., Fantastic Voyage, Batman,* and *Independence Day*; even comedies got into the act with *Spaceballs, Get Smart, Austin Powers,* and Woody Allen's *Sleeper*. That's a span of 40-plus years of SAGE in Hollywood. SAGE's whirring, flashing racks of iron became the obligatory standard for anyone contemplating the design of a backdrop for any sci-fi or even remotely futuristic movie or TV show. What a hoot it was for any bright boy at a movie theatre or in front of a TV set to witness Hollywood's best cavorting with their much-maligned machine: the "big boondoggle" in the Barta Building. SAGE became a computer hero! No telling how many times it averted disaster for the world or how many millions of lives it saved, all in pursuit of big-screen storytelling. SAGE's name never appeared in the film credits, but it was nevertheless up in lights on the silver screen. A much weightier venue, the 1996 presidential campaign, also used a SAGE backdrop. The then-famous computer with decades of screen appearances was added to lend some high-tech credibility to the *World News Tonight* stage set. ABC News carted in pieces to sit behind newscasters Peter Jennings and David Brinkley as they speculated on the Clinton–Dole returns. No one need venture into a Radio Shack to see the outcome of the digital revolution. All that was necessary was a comfy chair at home in front of the tube with a TV dinner most any night of most any week.

The Big Day

In 1950, "there was only a fraction of a megabyte of random access memory on planet Earth, and only part of it was working at any given time."[29] For those few souls trying to usher more of it into being, progress was frustrating and haltingly slow. In March of that year, Mina Rees and her ONR mates would journey to Cambridge[30] and unwittingly jeopardize about 4,000 bits of that precious resource. A practical mathematician and administrator, Rees' main intention in coming to the Barta Building was to wrest back under control what she and the ONR perceived as an extravagant budget. She had no particular malice toward memory; she gave it little thought, and thought even less about the far-reaching consequences of her actions. Sixteen hot bottles each with 256 bits of fragile memory intended for Whirlwind's racks awaited her arrival. The bright boys needed somehow to stop her.

In large measure the early progress of the electronic digital computer was dependent on the discovery and development of fast, reliable primary memory, what we today call RAM or random access memory. RAM is the vital, active memory that computers manipulate on the fly while in operation. RAM briefly holds binary text characters, symbols, and numerical values as the computer fetches and executes instructions. In 1950 electrostatic tube RAM was the fastest technology available. Whirlwind's first set of 16 electrostatic tubes had a paltry total of 4,096 bits of RAM (8 bits equals 1 byte). Eventually the bright boys would plump out primary memory to two banks of 16 electrostatic tubes for "a total storage capacity of 2048 sixteen-digit binary numbers,"[31] effectively doubling Whirlwind's primary memory.

Each tube would hold a single digit of the 16-digit binary number, and Whirlwind would make a sweep of all 16 digits simultaneously, taking 2 microseconds to add two of the binary numbers or 20 microseconds to multiply two of the binary numbers.[32] Everett's block diagrams group designed a 16-digit word length (5 instruction bits and 11 address bits)[33] for holding down the cost of the computer. Many mathematicians of the time, especially those from the ONR, considered the word length way too short to be useful,[34] although the first production IBM personal computer (1981) also used a 16-bit word length.[35]

Such infinitesimal amounts of memory—as well as such slow computation speeds—are nearly incomprehensible when compared 60 years later to a laptop computer that would handily toss about a gigabyte of RAM, the equivalent of a billion bytes (exactly: 1,073,741,824 bytes). Such a remarkable transformation from bits of memory to gigabytes of memory would not begin until 1970 when the Intel Corporation produced the 1103 silicon memory chip, which possessed a single kilobyte (1,024 bits) of RAM. In terms of physical size, it was on the order of comparing a magnum of champagne to a postage stamp. And better still, there was virtually no heat; there were no electron guns firing beams through huge vacuums at holding plates, no counting the hours until each hot bottle burnt itself to extinction. However, between the bottles and the 1103, between the 1950s and 1970, there was a wasteland of potential alternatives to tube memory.

Serial memory from rotating magnetic drums was too slow as were acoustic waves flowing slowly—at the speed of sound—through mercury delay lines. The only interim solution was random access magnetic-core memory, which in March of 1950 existed only in the pages of Forrester's notebook. Profound change in the history of computing lay in Forrester's drawings and in his Deltamax tinkerings with magnetic donuts strung over a grid of wires. Rees' budget visit would threaten more than the ONR's cash flow to the bright boys; it was disaster in the making for electronics and computer technology. In view of the geopolitical events of the day, with the dread of holocaust, invading armies, revolutions, terrorist threats, and rampant fears of a world running amuck, halting any progress going on in the Barta Building was an ill-conceived roadblock to national security.

"Two conferences took place on March 6, 1950, one in the morning and one in the afternoon."[36] Into the Barta Building first thing Monday morning flowed the MIT contingent led by Provost Julius Stratton, with Mina Rees heading up the ONR group. Neither Forrester nor Valley nor any of the bright boys attended the morning session. It was billed as a "policy meeting," which was MIT's way of softening up the scene for both the entrance of a white knight and amicable surcease for the ONR's wallet. It was all accomplished with a smile over coffee, without any prancing from a white knight or wiseass researchers gloating in their cups. The ONR was far from blindsided by the meeting. ADSEC's activities, ongoing from January 20th, were well known, as were those of the Ad Hoc Committee on Air Defense, as were Marchetti's inquiries to the ONR about Whirlwind's availability.[37] Each side knew the outcome well before the meeting began; all that was really necessary was a formal face-to-face and some polite, academic-style niceties.

The afternoon session would be ideal for a gloat or two, seeing that Rees' Charles V. L. Smith, who had savagely ripped Forrester's budget projections, was there for the changing of the guard. Forrester and Valley were ready for the event; Valley had Air Force authority to spend and a $500,000 opening offer to throw on the table. It was an eye-opener, and very quickly "all agreed that this would be an excellent solution to the situation."[38] Valley said later, "I could have gotten much more, but that's all they seemed to want at the moment."[39] He explained that his needs for Whirlwind were to run experiments whereby ADSEC could test out the feasibility of sending radar signals to a digital computer. The hope was that such an arrangement would give the Air Force the ability to quickly scan, track, and intercept multiple aircraft simultaneously. It was agreed that both the Navy and MIT would still have access to Whirlwind for their computing needs; the Watson Laboratories' air traffic control contract, including the services of Welchman and his crew, would be transferred to ADSEC; and the ONR's $300,000 for the upcoming fiscal year would be forthcoming. Including the remaining budget for the existing fiscal year, Forrester suddenly had continued financing of over $1 million, which in the desperate struggle to keep Whirlwind alive put his survival skills somewhere between those of the enigmatic Harry Lime and the crooning of Ezio Pinza. Rees was more succinct in the reason for the bright boys' close shave: "They were lucky."[40]

The white knight's journey to the March meeting was strangely serendipitous and convoluted, especially for an anti-nuke physics professor, comfortably tenured, easing his way into a stretch of benign course instruction with a little bit of cosmic ray research on the side. In 1950, George Valley was 37 years old with a family and a new home in Belmont. Gone were his Rad Lab days of flying to London in a bomb bay and winging home in a flying boat by way of Brazil. He had done his bit. He was not personally challenged at breathing life into a monster computer as Forrester and his bright boys were. He did not need to shoulder the weight of attacking TU-4s like Fairchild and Saville. He didn't have the all-consuming drive for national security as did Bob Robertson. He could have just waved off Robertson's taxi with, "No time now, I'll call you later." He could have escaped into the pursuit of postwar normalcy just like millions of others. But there he was in the Barta Building with millions of dollars to spend and cart blanche to spend it. Eyes from everywhere would soon be on him; his reputation would be on the line; he was in deep, right up to his chin, attempting to solve a problem that just might have no solution. And if the ambitious calling he had volunteered for wasn't enough of an undertaking, naysayers added more. "When word spread . . . that I had gotten in bed with Whirlwind, a number of busybodies warned me that it was a grave error . . . and I found myself snubbed in the halls of MIT by a personage very high in its administration."[41] Valley, as Forrester had done before him, had thrust himself into an awkward situation where, whatever the outcome good or bad, he was there for its duration.

George Valley was born in New York City on September 5, 1913. At 22 he earned an undergraduate degree in physics at MIT, and in 1939 a PhD in nuclear physics from the University of Rochester, New York, which was followed by a turn as a lens designer at Bausch and Lomb's facility in Rochester. He found himself back in Cambridge later in 1939 as a Harvard Research Fellow, and a year later trooped in with the gang as the Rad Lab was getting underway. He stayed until the Rad Lab mustered everyone out at war's end. During his stint there, he developed the H2X all-weather radar bombsight, which proved vital during the particularly foul weather years in Europe of 1943–1944. In 1947, he put in a year as editor of the *Radiation Laboratory Technical Series*, which turned out to be a huge asset for postwar electronics research. And except for his ongoing membership in the von Karman-led Scientific Advisory Board, the advent of 1949 seemed to be offering up to him some respite to pursue teaching and the good life. Then came Bob Robertson's, "Hello, George."

First Encounters

George Valley had major mojo. He was either outrageously fortunate or had the market cornered in lucky rabbits' feet or was the object of divine intervention. As the leader of a modestly sized air defense project that was about to rocket into the realm of the gargantuan, he never once along the way (from Whirlwind to SAGE, 1950–1958) stumbled into a project-killing event or person, and never fell victim to dead-ends, huge losses in time, lack of money, shoddy engineering, or just bad technology. That's not to say that he wasn't constantly menaced by misfortune; he was, but he somehow adroitly sidestepped every threat. He not only avoided bad fortune but also seems to have had an unerring knack for falling into good fortune. As luck or rabbits' feet or divine intervention would have it, he experienced a series of near miraculous first encounters with people and events that always proved beneficial and shoved the project forward. Some of it, as Bob Wieser saw it, was due to the quality of people around him and their dogged belief in their ability to pull off the seemingly impossible. Much also had to do with Valley himself, his inner fortitude, his courage at taking on an ostensibly hopeless job that he could have easily avoided, the experience he gained from his war years at the Radiation Laboratory, and the wisdom gained amid the desperate struggle that was the Chain Home and the Blitz. There was something about Valley, something ineffable, something special that always seemed to make things go the right way at near precisely the right time.

Valley's string of first encounters was too remarkable to be mere coincidence. The taxi ride at the Pentagon with Robertson was pure serendipity, but it jump-started everything else and set up the encounters that followed. Valley's encounter with the Lash Up system brought him to the doors of the AFCRL's John Marchetti, the Father of American Radar, who (according to Truman's science expert, William Golden) had the best military research facility in the country, loaded with good researchers like John Harrington, who would soon become vital in creating the SAGE network. His quick analysis of Lash Up and his out-of-the-blue phone call to von Karman were both timely and right on the money. That phone call then prompted Valley's next encounter, which was with Muir Fairchild, the exact right military mind and temperament for the air defense project, who got Vandenberg to quickly OK Valley's plans and Congress to rustle up more than enough money to begin. Through Fairchild and the Scientific Advisory Board (SAB), Valley next encountered the brilliant, hand-picked aerodynamicists for his Valley Committee.[42] The Valley Committee meetings intersected him with Gordon Saville (by then a Major General), the country's premier air defense expert. Valley next bumped into Jerry Wiesner in the corridors of MIT, a meeting that produced a very fortunate, January 1950 luncheon with Jay Forrester. That bread-breaking encounter was followed up by a visit to a "working" Whirlwind, which got Valley his Air Force money from Fairchild and control of the Barta Building from Rees. It was a remarkable seven months with nary a hitch, stumble, or false move along the way. And that was only from September of 1949 until March of 1950; Valley still had an additional seven years' more of first encounters ahead of him. He would not disappoint.

The Air Defense Engineering Committee (ADSEC) cranked up for business on January 20, 1950, six weeks before Valley's big day at the Barta Building, and a full five months before that watershed moment in preparedness, the outbreak of the Korean War. It was a Friday; ADSEC would continue meeting on Fridays for the next two years. Very quickly, the ADSEC acronym was dropped in favor of the Valley Committee: it was easier to say and more direct, so it stuck. Besides, the looming figure of George Valley was associated with every aspect of the air defense group and made his name a natural and convenient reference. Much like the wartime Sunday Soviets at the Grosvenor Hotel, these Friday Soviets similarly consisted of civilian scientists, engineers, and the military informally yet formally mingling together all bent on the very same challenge: air defense. Accommodations were far less posh: the Valley Committee met in a rather spartan room at the AFCRL on Albany Street. But it was a handy place to get to for all involved and, better still, was located at the spiritual center of the newly emergent digital electronics. As a practical matter, the ADSEC site was also selected so as to be near a large city, "so the final product could be used to defend that city as well as to serve as a model for the other installations."[43]

Valley further suggested that his fellow committee members be drawn from the fields of physics, electronics, aerodynamics, and guided missiles. ADSEC had nine permanent members,[44] four of whom were SAB affiliated and also taught at MIT—Guy Stever and Charles Draper were aeronautical engineers with expertise in guided missiles; Henry Houghton was head of meteorology; and William Hawthorn was a mechanical engineer and renowned expert on jet engines. Aerodynamicist Allan Donovan flew his own plane in every Friday from the Cornell Aeronautical Laboratory. Physicist and wartime blind-landing researcher, George Comstock, vice president of Airborne Instruments Laboratory, rounded out the group together with Valley, Marchetti, and Air Force liaison Major Richard Cella. Any week could see an ADSEC meeting swell with any number of invited guests or a specialist lecturer or two.

Valley and Marchetti set out an intensive weekly work schedule for the Committee. It began innocently enough with quick reviews of the air defenses of the Royal Air Force and the *Luftwaffe* and briefings from the United States Air Force on the current US capability. Most of the introductory topics were casual, get acquainted-type ice breakers for the nine-member group. Subsequent topics were more to the point: a Raytheon ground radar proposal, a Douglas Aircraft proposal for a supersonic jet interceptor, new work in X- and L-band radar, airborne radar, airborne analog computers, a visit to Air Force Headquarters for a Gordon Saville air defense briefing, an investigation of the potential of unguided rockets, the use of telephone lines to forward radar information, and the use of a computer to gather, store, and distribute radar position data.

For the telephone–computer connection, the Valley Committee meeting of February 17th started with Marchetti briefing on the use of continuous-wave radars tied to a computer, which was followed by Jay Forrester pitching Whirlwind and its "10,000 arithmetic computations per second."[45] It was easy for the members to see the simple logic in the use of a computer. "There was no conceivable way in which human radar operators could be employed to make these calculations for hundreds of aircraft as detected from such a large number of radars."[46] The feasibility of using telephone lines to send position data to a computer was later presented by Western Electric's Claire (Hap) Halligan (future first president of the MITRE Corporation). Western Electric's expertise was based on its astounding, 1942 installation of the Pentagon's telephone service (the Pentagon, the largest office building in the world, was built in 16 months), and was the communications nerve center used to direct the entire defense effort for all of World War II. It had the world's largest private branch exchange (PBX) with 13,000 lines of dial PBX equipment, 125 operator positions, and 100,000 miles of telephone cable. Wiring a continent full of radars to computers was a job that Western Electric seemed ideally suited to take on. Very early on in the life of the Valley Committee, Valley and Marchetti were hammering home their conviction for radar signals over telephone lines to an electronic digital computer. It found a receptive audience with the Valley Committee. They voted to include a line item in the budget to rent Whirlwind.

Whirlwind's salvation was assured by February, and Forrester, along with many of his fellow bright boys, became regular attendees at Committee meetings. It was a strangely new presence for otherwise traditional air defense meetings between radar engineers and aerodynamicists: electronic digital computer guys were now in the mix bringing along their own argot of clock speeds, central processors, memory banks, programming, and architectures. Strange as the initial intrusion may have seemed, there was, however, enough of a commonality among the disciplines to see the inescapable logic and inevitability of engineering it all together. The Valley Committee was for the first time contemplating the building of an intricate, real-time, electronic information system. This was not a railroad system, a waterway system, an electric power system, or even the telephone. This was to be the conscious and systematic manipulation of electrons to pass digital information among and through a series of complex electronic machines in order to form a single, dynamic system. This was something new. These men were not thinking in terms of radar only or computers only or data transmission only; they were thinking about an entire interconnected system: systems thinking. Systems thinking preceded the actual designing of the interconnected dynamic system: systems engineering. Simon Ramo, an early systems engineering advocate and chief engineer of the Air Force's Atlas missile program from the early 1950s, best explained it: the function of systems engineering is to "integrate the specialized separate pieces of a complex of apparatus and people—the system—into a harmonious ensemble that optimally achieves the desired end."[47]

Marchetti went so far as to build a scale model of the proposed radar-to-computer system; and the Committee voted for June of 1950 "to make an early test by hooking up one of Marchetti's radars at Bedford to Whirlwind, using Harrington's phone-line apparatus."[48] "At the time, Whirlwind had no electrostatic storage," recalled Bob Wieser, who, along with a group of graduate students that included David Israel, Robert Walquist, Jack Arnow, and Howard Kirschner, switched from working on air traffic control with Whirlwind to air defense. "Random access memory was five flip-flop registers and 32 toggle switch registers that could be read by the machine. We got the radar data into the machine and displayed."[49] The radar echoes sent from Hanscom Field's MEW Hill radar to Whirlwind in the Barta Building would prove to be not only an important test for air defense but also a singular moment in the birth of Information Technology: information theory, cybernetics, and digital computing all converged with the first packets of digital information sent from Bedford radar over telephone lines to Cambridge.[50]

It was the first step in a digital journey that would culminate when Whirlwind computer descendants, the TX-2 and IBM's AN/FSQ-32 (based on AN/FSQ-7 and AN/FSQ-8 technology), were networked together by Larry Roberts on the East Coast and Tom Marill at SDC on the West Coast.[51] Just beyond that piece of networking lay the shores of the Internet. In fact, Roberts is often called the Father of ARPANET, which was the predecessor of the Internet. It would take more than a decade of work to reach the watershed event of transcontinental computer-to-computer networking. The Bedford-to-Whirlwind detect and intercept experiment in

April of 1951 came first and would be scaled up to a larger experiment called the Cape Cod System in 1953, which scaled up even larger, becoming SAGE in 1956, and then to the first mammoth SAGE installation at McGuire AFB in 1958.

Forrester and Everett had theorized on such system's ideas in their L-1 and L-2 laboratory reports to the Navy in 1947. Louis Ridenour (SAB member and USAF chief scientist) hit close to the mark, authoring in 1947 one of the famous Radiation Laboratory textbooks, *Radar System Engineering*. A case for early concepts on systems engineering can even be made for the original thinking found in the master's theses of Perry Crawford (1942) and David Israel (1951). However, nothing previous to the Valley Committee approached the scope of anything like SAGE systems thinking. It is interesting to note that the keen-eyed word hounds at the *Oxford English Dictionary* took serious note by including terms like "systems engineering," "systems analysis," and "general systems theory" in *OED* editions after 1950. Later, in 1957, Harry Goode, along with coauthor Robert Machol, would be credited with authorship for the early bible on systems engineering, *Systems Engineering: An Introduction to the Design of Large-scale Systems*.[52]

The Friday, March 17th Valley Committee meeting returned to the nitty-gritty of radar, specifically, "the serious limitations of the new generation of ground radars."[53] Over-water ducting of radio waves, situation displays, and data handling problems were also covered. Forrester attended, weekly getting more familiar with the overall system in which his machine would play the central role. The very next week, March 24th, Forrester returned with five of his fellow bright boys; 15 other visiting specialists also joined the meeting as guests. Forrester discussed the "problem of combining data from three or more radars, and mentioned the need for storage tubes; Harrington discussed terminal equipment for pulse radars that were to be used with a computer."[54] Systems engineering the air defense pieces together into a single, dynamic entity was beginning to pick up speed.

In the spring of 1950, to intensify the hunt for low-altitude radar coverage, Valley began holding additional "Thursday-evening radar seminars to which mathematicians, engineers, and physicists from industry as well as universities and Air Force installations were invited."[55] Picking up speed as well were the ranks of the attendees—the spartan room at the AFCRL was starting to swell with curiosity, interest, and popularity. Valley wrote that ADSEC's "renown began to grow" and that in July he was invited to a Pentagon briefing along with 21 generals and the Secretary of the Air Force.[56] Barely five months after his Barta Building meeting with Mina Rees, Valley experienced a first encounter with the totally unexpected: fame.

The Competition

The landscape for electronic digital computing in the early 1950s seemed a scary place in which to venture either one's money or career.[57] There really was not much of it going on anywhere, except for a dozen projects that the military had ordered and a sliver of emerging interest from private industry. It wasn't taught in college and mothers weren't urging their sons to take up the cause of the digerati.

In actuality, the early 1950s would prove to be the best of times in which to take a chance on a digital future. The bright boys were quite convinced of it. They had endured and come through; others could as well. They felt sure that theirs was a technology that had nowhere to go but up. "Digital computation is a new and growing field and a young and rapidly developing profession," they wrote in their laboratory's brochure.[58] They were sky high over their own newfound calling's sudden success, which was made even sweeter as Whirlwind became quite the tourist attraction, generating in a single year (1953) 990 curiosity seekers. Some 95 industrial companies sent representatives, as did 40 government organizations, 35 academic institutions, and 23 countries.[59] It was a far cry from the accusingly suspicious faces that had stalked them from 1948 to 1950. The bright boys were sure about the road to success, insisting, "It is not the machines but the people that are the top asset of a laboratory." And, they felt that a digital apprenticeship might well be the key: "Digital computation is a new field which has been successfully entered by men of good ability having no prior experience in the work."[60]

Roger Sisson was a prime example of the allure and transformative effect of the Whirlwind experience. In 1948, this 22-year-old kid from Brookline, Massachusetts, with a fresh degree in electrical engineering from MIT was inspired by a Jay Forrester talk on computers, and was hooked. That September he found himself in room 212 of the Barta Building, sharing space with other bright boys, Charles Corderman, Norman Zimbel, and Harrison Rowe, just across the hallway from the great one, Forrester himself. He worked on cathode ray tube displays, and then together with Alfred Susskind, another kid from downstairs in room 124, did his master's thesis: one wrote about analog to digital conversion, the other about digital to analog. By 1950, the new Whirlwind convert found himself on the West

Coast in a doctoral program at UCLA and employed at the recently-formed Electronic Engineering Company of California, EECO. Sisson was but one of the scores of brilliant, youthful acolytes drawn to the humming of the great machine and to mingle with its band of brothers at 211 Massachusetts Avenue. Many, just like Sisson, would issue forth from that seminal training experience to form the vanguard of a new breed of computer engineer. Later, SAGE would do likewise, but by the hundreds.

However, patrons or benefactors were necessary in order to survive until the field quickened. Electronic digital computing had three types of early patrons: the military, universities, and industry. Saul Rosen's 1969 retrospective "Electronic Computers: A Historical Survey" traces the impact or lack thereof of all three.[61]

Rosen's perspective is a uniquely balanced one. He was a mathematician with an engineer's sense of machinery and circuits, spent the war in the Army's Signal Corps, was a computer builder, a founder of the Purdue University computing laboratory and computer engineering program, and worked for years within the early electronics industry as it strove to push machines out into the marketplace to make a profit. Rosen had lived it all. With a PhD in mathematics from the University of Pennsylvania, Rosen worked the teaching circuit from the University of Delaware to Drexel to the University of California to Wayne State to Purdue. He labored for industry giants Burroughs, ElectroData Corporation, and Philco. With Philco he was the chief engineer on the world's first commercial, transistorized computer, the Transac S-2000. His famous 1967 book, *Programming Systems and Languages*, was a rite of passage in computing education. And for many years Rosen was an editor of the *Annals of the History of Computing*. Rosen wrote as an eyewitness to the origins of computing, seemingly to every moment of its unfolding.

A concise 23 pages, "Electronic Computers: A Historical Survey" is an insightful insider's diary on the progress of electronic computing baby stepping its way from 1946 to 1966. Remarkable in Rosen's comparison of the technology's machines, their makers, and patrons is how far ahead the bright boys and Whirlwind were from everyone else outside the doors of the Barta Building. Rosen does not deliberately exalt them as being kings of the digerati, but they nevertheless rise to that level as he describes the computing landscape around them.

Rosen begins with the digital font, ENIAC, at its digital epicenter, Philadelphia. In 1950, Philadelphia was trying hard to become America's first Silicon Valley without the silicon. The great ENIAC was purring away for the Army at the Aberdeen Proving Grounds while its younger brother EDVAC was still coming to life at the University of Pennsylvania. Not far away, and also pushing into existence, was the soon-to-be-famous UNIVAC at the Mauchly-Eckert works, where it was heading toward a rollout date in March 1951. Preceding their UNIVAC by two years was the Mauchly-Eckert BINAC, the first operational stored program computer in the United States. While BINAC, a mercury delay line machine, worked well in factory tests, it did not ship well and never worked for the customer, Northrop Aircraft, 3,000 miles away in Hawthorn, California. "The poor engineers were constantly working to get BINAC to run," recalled Roger Mills, eyewitness to the struggle. "One side would be running while they worked on the other side. The two sides never worked together as long as I was there."[62] The BINAC's avowed purpose in life was to be part of the guidance system for the top-secret Snark missile; its enormous size precluded its riding copilot for the missile. However, its computing capabilities opened more than a few eyes very wide at Northrop. "The most successful run we made was on a deicing problem for an airplane. Two operators on electric calculators worked for 6 months computing steps in resolving the differential equations. The BINAC did these steps and completed the calculations in 15 minutes."[63] Fifteen minutes wasn't real-time computing but in comparison to six months, it was lightning-like and very pleasing on the fingers.

Early on, Mauchly and Eckert not only produced leading-edge technology but also, and thankfully so for them, produced a financial backer to cover their inept cost accounting that allowed UNIVACs to be woefully under priced at $250,000 apiece together with promises for delivery that were utterly impossible to fulfill.[64] Harry Straus, founder of the American Totalisator Company, which made the "Tote Boards" that are fixtures at nearly every thoroughbred race track in the country, invested to the tune of $500,000, becoming chairman of Mauchly and Eckert's company. Computers were a natural add-on to his electromechanical systems for calculating odds, dispensing tickets, and displaying payouts on horse races. Unfortunately, Straus died in an airplane crash in 1949 and American Totalisator's board quickly abandoned Mauchly and Eckert. With loans coming due in January 1950, they hustled their company around the city looking for a buyer: Burroughs and Philco passed; Remington Rand had interest.[65] Hats in hand, Mauchly and Eckert also met with IBM's Tom Watson Sr. to try to work out a future. Watson, who would have nothing to do with anything that did not relate to the punch card business, declined their offer.[66] Watson rejected computer technology that was more than enough to give IBM leadership in electronic computing for years to come. But it was not the first time that his neophobia cost his company a leg up in technology. He had also flatly refused patent attorney Chester Carlson's 1947 offer on an electrophotographic printing device that Carlson later sold to the Haloid Photographic Company, which, in turn, changed its name to Haloid Xerox, and still later to the Xerox Corporation, as the enormous success of xerography began to take hold.[67]

Mauchly and Eckert next hightailed it to Florida for a meeting on Jim Rand's yacht, where the owner of Remington Rand—another Philadelphia powerhouse—bought out their floundering company and the rights to UNIVAC, America's first commercial computer, on February 1, 1950.[68] For under $1 million Rand landed the best data processing computer in the country, together with $1.2 million in existing UNIVAC contracts from the likes of the Prudential Insurance Company, AC Nielsen, and the United States Census Bureau.[69] The very next year—1951—exhorted along by Watson's son, Tom Jr., IBM would begin a crash course to produce its own electronic digital computer;[70] playing catch-up all the while with a machine that it had let slip away. It is rumored that Watson Sr. nearly blew a gasket when Walter Cronkite, anchoring CBS News, televised the results of UNIVAC predicting the Eisenhower upset in the 1952 presidential elections. The emerging significance of that event was lost on Cronkite. He recalled later that, "I don't think that any of us saw the long shadow in the newsroom at all."[71] Watson had. And he accorded even greater notice to big IBM customer General Electric buying a UNIVAC to do its payroll.[72] Remington Rand would strike Big Blue again in December of 1951 when it bought ERA and the rights to its Atlas computer. IBM's Werner Buchholz remembers it as a time of drop everything and full steam ahead to push out a UNIVAC competitor: "On the 701 we started with the so called, Von Neumann machine, the IAS machine at Princeton, as a starting point and of course Whirlwind."[73] Watson also had in his stable of engineers Nat Rochester, the guy who built Whirlwind's arithmetic unit and had taken Bob Everett's course on electronic digital computing.[74]

Fortunately for IBM the ghastly situation was more than salved in September 1952 when it encountered Project High, which was the movement to convert Whirlwind into the AN/FSQ-7.[75] Project High got its name from a necktie factory on High Street in Poughkeepsie where IBMers and the bright boys met for discussions. IBM's management understood well the importance of getting the AN/FSQ-7 deal: "IBM will be recognized as the undisputed leader in the large-scale, high speed, general purpose digital computer field. If a competitor were performing on this contract, that competitor might gain enough advantage to force IBM into a relatively secondary position."[76] By April 1953, Watson's long-awaited response materialized: IBM rolled out a computer, publicly announcing the arrival of its IBM 701 Electronic Data Processing Machine. IBM was finally in the game. And when it was selected over Remington Rand and Burroughs to produce all of the AN/FSQ-7s, IBM was *really* in the game.

The Burroughs Adding Machine Company tried to bull its way into computing when it nipped in from Detroit to the Philadelphia suburb of Paoli in 1948 to open a research facility. Burroughs's vice president charged with starting up its computer division was Irven Travis—the former supervisor of research for the Moore School who tried to force Mauchly and Eckert to "sign a patent release or else" on their ENIAC. Burroughs unwisely passed on the Mauchly-Eckert business; then labored on for years afterward, finally producing in 1956 the lowly E101, which prompted the wisecrack about "the mountain that had labored to produce a mouse."[77] Acquisitions of technology from Control Instrument Corporation and then the ElectroData Corporation enabled Burroughs to produce a data processing machine, the BEAM IV, and later, the 220. Both were far from being barnburners in electronic computation.

Soon to fall under the sway of Philadelphia's computing hegemony was ERA (Engineering Research Associates) of St. Paul, Minnesota. In 1950 ERA was still in the glider factory (originally an old foundry for the American Radiator Company) working on top-secret cryptologic machines for the Navy's crypto group CSAW (Communications Supplementary Activity, Washington). Their work was so secret that the plant was declared a Navy Reserve base and armed guards were posted at the entrance. ERA perfected its plundered German magnetic drum technology, providing the Navy beginning in 1947 with a series of magnetic drums for cryptologic storage.[78] ERA even made a deal with IBM to share its drum expertise, when IBM's own research efforts in the technology seemed ready to falter. Magnetic drums led to computing soon after CSAW's James Pendergrass returned from the Moore School lectures. His eyes were opened there and electronic computing "was what we wanted as a logic machine for enciphering and cryptanalysis."[79] His paper to the Office of Naval Research "was a successful tool in selling the basic idea within the Navy that programmable computers were the wave of the future."[80] Mina Rees and company must somehow have missed or overlooked for three years the Pendergrass paper circulating around the ONR beginning in October of 1946.

The Navy issued task order 13 to ERA in 1947 to build a high-speed, cryptologic computer. Task 13, code name Atlas, was for secrecy purposes not referred to as a computer but rather as an "analytical engine."[81] Atlas was hurriedly built and then delivered to CSAW's headquarters on Nebraska Avenue in Washington, DC in December 1950 (in 1952 CSAW would become part of the National Security Agency). ERA's operating budget, including $1.5 million for their very first year of operation, was secretly buried in the budget for the Navy's Bureau of Ships. It was large amounts of such slippery-looking financing from strange budget sources like the Bureau of Ships that was a frequent Forrester rant to his own tight-pursed Navy patrons. Whirlwind's expenses, argued Forrester, were less than ERA's and, besides, Whirlwind was a better machine.

Even before the completion of task order 13, CSAW asked ERA to start on the design of another more powerful machine. "This project was called Task 29, and the computer was designated the Atlas II. Work got underway in 1950, and to supplement its staff ERA hired many of the 1950 and 1951 electrical engineering graduates from the University of Minnesota. One of them was Seymour Cray, who in later years became famous for his work on supercomputers at Control Data and his own companies."[82] Forrester was not alone in his observations of ERA. Columnist Drew Pearson's "Washington Merry-Go-Round" newspaper column had the very same rant against ERA in August of 1950.[83] What the column called a "highly secret, multimillion-dollar contract" where Naval officers suddenly disappear and then reappear as "highly salaried vice presidents of the company" was in need of some government scrutiny.[84] Federal auditors were soon into ERA's accounts in the Navy's procurement offices,[85] and ERA's gush of funny money trickled to a close. The end drew near when James Rand appeared. By December of 1951, Rand's offer of $1.7 million was accepted, and with it, Remington Rand suddenly was the country's biggest and best computer company.

A Philadelphia digital outpost was located 42 miles to the north at the Institute for Advanced Study (IAS) in Princeton, New Jersey. Von Neumann and his recruits of ex-ENIAC builders had dashed off for Princeton just after the Moore School lectures. They began putting together their mathematician's dream machine using, as had Whirlwind, electrostatic storage tubes. After struggling with storage tube delays, they finally got it fully operational in June 10, 1952, two years after Whirlwind. Von Neumann was quite open with the design of the IAS—he, Burks, and Goldstine, beginning as early as 1947, freely circulated all of their plans.[86] The result, a boon to early computer building, was a spate of copycat IAS machines popping up seemingly everywhere: the Johnniac (named for Neumann) at the RAND Corporation, the Ordvac and Illiac at the University of Illinois, the Maniac at Los Alamos, and the Weizac at the Weizman Institute in Israel. Rosen would use these same plans to build a transistorized version of von Neumann's IAS machine.[87]

At the corner of C and Tioga Streets, the Philco Corporation (which began its business life as the Philadelphia Storage Battery Company) was not overly keen on the electronic computer business, probably because it was the largest and most well-known radio manufacturer in the world. Philco radios were quite literally in every corner of the globe. What Philco was real keen about was building the first transistor radio. Failing that, they settled instead for building the world's first transistorized computer. Invented in 1948, "the transistor was expected to become the key to revolutionary advances in computer technology." It wasn't and would not happen for another six years. "Switching speeds were relatively slow; it was difficult to produce transistors with uniform characteristics."[88] The transistor was shuffled off to serve the hearing aid and radio business, and with radios it met head-on with the radio king, Philco.

Getting bulky tubes and heavy batteries out of portable radios was a major effort with radio makers of the 1950s, and to be the first to put out a transistor radio would be a tremendous market coup. Philco did not produce the first transistor radio: the Regency TR-1 was the first, and that was not until 1955. Worse still, the Regency wasn't very good. Poor technical quality with a $49.95 price tag, which in 1955 dollars was nearly a week's pay for many people, prompted *Consumer Reports* to warn in its April 1955 edition: "The consumer who has been waiting for transistor radios to appear would do well to wait for further developments before buying."

The audio quality, especially of the human voice, left much to be desired. The hunt for a better transistor was on, and Philco research made the necessary breakthrough in 1954 with the surface barrier transistor. The Philco transistor was so good that it made high-speed computing a reality. A year later the National Security Agency (NSA) sought out Philco to build a transistorized version of ERA's Atlas computer.[89] Project SOLO, as it was called, pushed Philco into the computer business, and the resulting Transac S-1000 became the first transistorized computer. It was another computer first for Philadelphia. But, of course, as a top-secret NSA computer, it was not for the world to see. The first transistorized computer for public consumption popping onto the Philadelphia computer scene would be the Rosen-led team's Transac S-2000 in 1957.[90]

Electronic computing's first generation saw a host of other wannabe electronic computer manufacturers trying unsuccessfully to nudge their way in with the big boys of the young industry. Some would fail early and then attempt reentry as the surface barrier and diffusion transistors opened up the technology's second generation. Some would never return. Some would fail even with transistors. Some would be swallowed whole and then spat out again wearing a new corporate logo. Then too, some would come out of nowhere to become the Cinderella stories of the decade.

Raytheon lumbered onto the scene producing the very tardy RAYDAC (originally Raytheon's Project Hurricane Computer) of which only one was sold to the Navy. Raytheon then got together with Honeywell to form the Datamatic Corporation, from which Raytheon soon bowed out, leaving a Honeywell Datamatic division to produce a grand-sized dud called the DATAMATIC 1000. Radio Corporation of America (RCA) contributed its BIZMAC to what Rosen terms another of the "interesting failures" of the first generation.[91] The king of the keyboards, the venerable Underwood Typewriter Company, toe dipped into electronics, buying the Elecom Computer Company and setting up shop in Garden City, Long Island. Marketing their Elecom 120 as "An electronic computing system any business can afford," and going after the small to medium business market where Underwood typewriters were well known, was smart thinking. "Performs in one hour as much work as six people can do in a week on desk calculators," touted their 1953 product brochure.[92] Underwood, however, floundered into severe financial problems that forced its sale to Olivetti three years later.

Lack of financial staying power was the bane of many savvy marketing campaigns and brilliant technical innovations. A year after BINAC entranced Northrop, five engineers left to rush into the biz on their own. In a two-room office over a knit shop in Manhattan Beach, the West Coast chance takers started Computer Research Corporation (CRC) in May of 1950. One of their creations, the CRC 102, grabbed lots of attention and even got the Air Force Cambridge Research Laboratory to reach for its checkbook. But once again, another startup could not sustain itself and by 1953 CRC was owned by National Cash Register. One of CRC's founders, Richard Sprague, echoed the familiar lament of the day, "many computers sold and built, but no deliveries and no income. Internal squabbles, coupled with the death of a key financial backer."[93]

West Coast Harry Huskey seemed to be all the rage in Los Angeles County. The former ENIAC builder was not only at UCLA putting together the SWAC for the National Bureau of Standards but was also over at Pasadena's Consolidated Engineering Corporation (CEC) educating the staff through evening seminars on how to build their machine, eventually called the CEC-201. CEC spun off its computer division as the ElectroData Corporation, producing the Datatron in 1953 before being scooped up by Burroughs in 1956. Huskey next appeared on Arbor Vitae Street in Los Angeles at Bendix Aviation's Bendix Computer Division, where he built the refrigerator-size G-15 computer. Inexpensive for what was then around, the G-15 could be had for $14,900 or leased for $985 a month. Huskey even squeezed in an appearance on Groucho Marx's hit radio show, *You Bet Your Life* (moved to TV in 1951), where he gave as his occupation: designer of electronic brains. He most surely was.

In nearby Glendale there was no sign of Harry Huskey because another ENIAC veteran, Stan Frankel, prowled the premises of Librascope, a division of aircraft equipment maker General Precision, Inc. Frankel's eventual creation, the 740-pound Librascope LGP-30, of which a total of seven were made in 1956, was a desk-size machine that retailed for $39,600 and rented for $1,100 a month. Southeast across the desert and into Arizona the late-blooming General Electric (GE) was just setting up its computer division in 1956, headed by Homer (Barney) Oldfield.

As many as 26 private companies tried to get in on the early electronic computer industry. Most didn't make it. Even IBM had a tough time of it, but had monumental resources to wait for the right opportunity to appear. Few computer customers other than the federal government, large banks, insurers, and huge industrial companies could pony up the $1 million to $3 million in cash necessary to purchase one of the big, early computers. IBM's angle from the beginning, as it had been with their card tabulating business, was to lease their machines. Many more companies felt better about $16,000 on a monthly lease than blowing the corporate treasury on an extravagant outright purchase. To manufacture computers and then to lease them required enormous upfront capital to hang in there waiting until the lease revenue began to flow.

IBM had the bucks. With its 701 already in the marketplace in 1953, IBM announced the coming of its competition killer the very same year, rolling out its first 650 business machine in 1954. It was pricier than all the rest at $3,200 a month (just about a year's wages in 1953), but it was popular. IBM figured to manufacture 50 of the 650s but ended up installing nearly 2,000. "IBM's position in the punch card field was a tremendous advantage for the 650; for hundreds of business organizations it seemed to be the next natural step."[94] The 701 and the 650 proved to be a devastating one-two punch for other manufacturers, especially Remington-Rand. In addition, IBM had a highly skilled sales force that knew how to sell a lease. Remington Rand, on the other hand, although well heeled with a lineup of great machines, had an ineffective sales force made up mostly of technical types who were less convincing to potential customers.[95] By 1955, Remington Rand would find it necessary to merge with Sperry Gyroscope to form Sperry-Rand.

The mix of startups, mergers, and sellouts in the new computer industry roiled on from 1947 to 1957. In 1957, however, two Cinderella companies would emerge from the pack, each having waited out the storm of consolidations until the time was right. One Cinderella had been secure for nearly the entire previous decade working on government-supported projects. William Norris, a founding father at ERA, took a group of engineers from Sperry-Rand, including Seymour Cray, and bolted to form Control Data Corporation, setting up shop in an old warehouse down the road from Sperry-Rand in Minneapolis. The other Cinderella was led by Ken Olsen, an original bright boy and Whirlwind engineer, who moved into an old mill in Maynard, Massachusetts to form Digital Equipment Corporation (DEC). Olsen had built Whirlwind's Memory Test Computer (MTC) and had been assigned to help oversee IBM's production line for the SAGE AN/FSQ-7. It was no coincidence that when DEC shipped out the world's first modern minicomputer, it bore quite a striking family resemblance to Whirlwind and MTC technology. Control Data Corporation (CDC) quickly became a pioneering supercomputer developer, building the fastest computers in the world well into the 1960s and early 1970s. Digital Equipment Corporation eventually grew to be second only to IBM as the premier producer of business computing systems. By 1959 the world scene had nearly 2,500 digital computers humming away, which was an enormous leap from the fraction of a megabyte of random access memory of 1950. The end of the decade saw the computer industry in the United States winnowed down to what became known as Snow White and the Seven Dwarfs. IBM, of course, was Snow White, followed by the Dwarfs—Burroughs Corporation, NCR, Control Data, General Electric, Honeywell, RCA, and Sperry UNIVAC (the computer division of Sperry-Rand).

On February 17, 1950, as George Valley introduced Jay Forrester to his Valley Committee, the decade-long storm and stress of digital computing was barely astir. Forrester had before him a receptive, interested, even eager audience in the Valley Committee. It was an utterly pleasant change from openly contentious people with brows knitted in suspicion, as were

those from the ONR and so many others. The Committee was really pulling for Forrester and his machine to do the job at hand. There had not been
such fresh air around since the summer interlude that had been the Moore
School lectures. The promise of those days in Philadelphia had come to
pass; he and Bob Everett and the other bright boys had also come to pass.
Life was good again.

Maurice Wilkes said of them in his memoirs: "By the time I knew them
they were pillars of digital orthodoxy." Pillars, yet. What a nice, new chunk
of respectability had been hewn out of the former mountain of doubt. The
boondoggle boys became the bright boys to nearly everyone. Familiarity had
even abbreviated *electronic digital computer* to that of just *digital computer*;
and the death knell for analog machines was tolling ever more loudly. At
the front door of 211 Massachusetts Avenue a sign now read, "The Digital
Computer Laboratory," and Forrester was its director. They were ahead of
the pack in the field of digital computing and moving further ahead all the
time. While most others spent the 1950s with a head stuck in a machine
trying to crank it into existence, the bright boys were upright looking from
machine to machine . . . and pondering the possibilities of networks.

Chapter Eight
Into the Great Wide Open

The most profound technologies are those that disappear. They weave themselves into the fabric of everyday life until they are indistinguishable from it.

—Mark Weiser

Janus, the twin-faced Roman god of doorways and new beginnings, looks both back at the past and forward into the future. Janus was how Maurice Wilkes described Presper Eckert. "Facing two ways in history," said Wilkes, one way "the culmination of efforts of the past" while the other half "sets a new direction for the future." Wilkes was praising Eckert's intellect and industry for escorting the world through the doorway of digital computing with ENIAC. Much the same comparison can be made about Jay Forrester. In 1953 Forrester snapped the first frame of his magnetic-core memory into Whirlwind. It was a new beginning. With it, he forever slammed shut the door on unreliable tube memory that had been the bane of all early computers. From bane to balm, shaky computer memory quickly became a thing of the past. Programmers gained confidence that their meticulous routines would run properly; companies gained confidence that calculations of their all-important business data were sound.[1]

The enormity of such an event is near lost on modern computer users whose only concern with memory is having enough of it. IBM's 701 is a telling example of a great machine with a very big memory problem. As Saul Rosen tells it, "The electrostatic storage system on the IBM 701 was very unreliable . . . the mean time between memory failure at 701 installations was often less than 20 minutes."[2] The 701 outfitted with magnetic-core memory (renamed the 701M) was a huge improvement. Redesigned and resold as the IBM 704, it "was

Ivan Sutherland using Sketchpad program on TX-2 transistorized computer at Lincoln Labs, 1958. (Reprinted with permission of MIT Lincoln Laboratory, Lexington, Massachusetts.)

quite outstanding for its time and achieved a near monopoly for IBM in the large-scale scientific computer field."[3] In the 1950s, the advent of such trustworthy memory provoked a stampede for digital computing power.[4] Forrester's revolutionary invention was more than just a complete commercial success. It totally replaced skepticism with user confidence for an entire industry and would continue to do so for over 20 years thereafter.

When IBM received its first subcontract to convert Whirlwind technology into the AN/FSQ-7 in October of 1952, it reaped a windfall of bright-boy innovations to commercialize. IBM wasted little time in capitalizing on the technological advantage.[5] It not only produced the 704 but also began building the SABRE seat-reservation system for American Airlines, which became "the largest commercial real-time data processing system in the world."[6] And there in the thick of it all from the very beginning was none other than Forrester's digital friend and advisor, Perry Crawford. Crawford had retired from the Navy in 1952, and the very same year popped up at IBM as a member of its Product Planning group. He was the first to identify airline reservations as an ideal focus for SAGE technology.

Very much like SAGE with duplexed computers and a national grid of phone lines connecting nationwide control centers, SABRE connected 1,200 teletypes scattered all over the country to the airline's computing center north of New York City. Forty years later in his memoirs, Tom Watson Jr. acknowledged that the SAGE contract "enabled us to build highly automated factories ahead of everyone else, and to train thousands of new workers in electronics."[7] Bright-boy technology, modern production plants, and a skilled work force complemented both IBM's huge financial clout and its reputation from the tabulating machine business. It was easy for IBM to gaze into a mirror and see Snow White smiling back.

Remington Rand was unhappy that the bright boys passed it over in favor of IBM. Remington Rand also, and with good reason, was a bit surprised as well, especially seeing that its UNIVAC was the "it" machine of data processing, and that its acquisition of ERA's Atlas technology gave it a leg up in scientific computing. A memorandum from Forrester from November of 1952 explains the advantages of IBM over rival suitors Remington Rand, Raytheon, Bell Labs, and Sylvania.[8] The selection committee, made up of Jay Forrester, Robert Everett, Norman Taylor, and C. Robert Wieser, had a clear view of the computing landscape of 1952, and saw things much the same as their contemporary, Saul Rosen. IBM was the hands-down correct choice to entrust with the country's air defense computers. That IBM's Poughkeepsie plant was relatively close to Cambridge was also a good selling point. The proximity of IBM's branch office in Hartford, Connecticut also had considerable travel-time merit. It was there, following quickly on the heels of the earlier subcontract, that beginning June 24, 1953 groups of IBM and Whirlwind engineers rendezvoused to hash out AN/FSQ-7 design problems. At first testy and a bit combative, each side quickly settled in and came to understand and respect the other.[9] The Hartford meetings, known to both groups as Project Grind, were important first steps in coming to decisions about every system and subsystem of the huge project before them.[10]

George Valley and his Valley Committee were noticeably absent from the selection process, which says much about Valley's trust in Forrester and the bright boys' decision-making and managerial skills. Generally perceived as a bunch of cocky inventor types, the bright boys displayed considerable acumen in the very modern practice of outsourcing production. They rode herd on IBM as the latter first produced two experimental AN/FSQ-7 computers, called XD-1 and XD-2—one remaining with IBM as a working prototype for its engineers and the other shipped to Massachusetts as the bright boys' prototype. From these two machines, and any ongoing engineering advances performed on them, would come all the other AN/FSQ-7s.

For two sets of engineers separated by hundreds of miles of landscape, a good set of blueprints was critical. Everett's pioneering work in the annotated logical diagrams[11] of a computer's innards and connections, what today is called "block diagramming computer architecture," saved time, money, and anguish. In the fall of 1953, Forrester and Everett then set up a Systems Office run by bright boy John Jacobs to oversee design control and responsibility for the experimental XD computers through to the finished AN/FSQ-7 computer system. IBM followed their lead, creating its own three-person Engineering Design Office to interface with Jacobs' group.[12]

Necessary as well to the development process were accurate research documents to share between both teams of engineers. Peeling back the covers of Forrester's laboratory reveals an early penchant for painstakingly thorough documentation and communication of everything that went on, beginning in the early pre-Barta Building days and continuing onward into the AN/FSQ-7 and SAGE projects.[13] Since there were no existing publications on digital computing anywhere to be found, the bright boys published every scrap of their own research as they went along, even if distribution was limited only to themselves. There was the seminal Report R-127, and the forward-looking L-1, L-2, and L-3 reports, but they hardly scratched the surface of the volumes of paper produced. "There were progress reports, trip reports, conferences and conference reports, master's theses stemming out of work done in the laboratory, technical memoranda, and discussions with visiting experts from industry, academe, and government."[14]

Everything was categorized and filed either as an M-Note, C-Note, E-Note, or Bi-Weekly Report. This unorthodox, high level of communication remained consistent from the skeleton of shelves that was early Whirlwind to a continent filled with AN/FSQ-7s. All of it "word processed" by hand on clunky typewriters and then duplicated on mimeo machines.

Forrester even conducted weekly affairs called Friday afternoon teas where he'd speak informally one-on-one with individual researchers; these grew to group teas as time went on and the bright boys expanded their ranks. It was all about discipline in pursuit of reliability, which was the all-abiding goal to which all early computing ventures aspired, yet few ever achieved. Forrester and Everett demanded reliability. If an engineer worked himself into an electronic dead end, his own notebooks and the circulated knowledge of others were his only salvation. There was no running to a bookstore or library for the latest on digital circuit design and no cash to redo something over and over again until it worked. And their machine showed it: Whirlwind's reliability became legend in the industry.

Forrester's great advance was one of three momentous events from the early 1950s that propelled computer technology into the future. Another was the outbreak of the Korean War in June of 1950. Although Truman was still looking to cut defense spending by a half billion dollars as late as May of 1950, the North Koreans would change all that penny pinching in a hurry. The war sprung open the military's purse strings from $13.5 billion to $48 billion (July 1951), and sent the Department of Defense on a feeding frenzy for anything war-winning, especially electronics.[15] "The motivation was the fear that unless the United States engaged in a militarized containment of Soviet power, the Korean War could be a prelude to a much wider conflict with the USSR."[16]

Stalin's generals had drawn up the war plan and the North Koreans executed the plan perfectly, taking Seoul, the South Korean capital, in a matter of days. The danger of a wider war was far from an exaggeration. According to historian and Stalin biographer Edvard Radzinski, Stalin told his close Politburo buddies: "We have the opportunity to create a communist Europe but we have to hurry." By hurry, he meant while the US still had a demobilized miltary that had shrunk to pre-Pearl Harbor numbers and had its hands full of North Koreans. Before he could make good on his threat Stalin died on March 5, 1953 from a brain hemorrhage, or as Radzinski contends, was poisoned.

Curiously, four months later, on July 27, 1953, the war that Stalin started and the war that the United States felt sure was a harbinger of a wider conflict ended with a cease-fire agreement between the United Nations and

North Korea. Truman believed that Stalin was heading toward a showdown. The previous January in an interview with Carleton Kent of the *Washington Sun-Times*, Truman said of Stalin's Korean adventure: "It's the greatest error he made in his whole career. If he hadn't made that mistake, we'd have done what we did after World War I: completely disarm. And it would have been a cinch for him to take over the European nations, one by one."[17].

The Korean War, like all wars, especially modern ones, held great promise for technology. Technology would grow particularly prosperous in the 1950s serving an expanding military that was ramping up to spread itself all over the globe. As Charles Bohlen, a 1950s US Foreign Service officer in Soviet affairs points out in his *Witness to History, 1929–1969*, because of post-World War II demobilization, the US had military facilities only "in England, where we had transit privileges, and Saudi Arabia, where we had an airfield. As a result of our over interpretation of Communism's goals, we had by 1955 about 450 bases in thirty-six countries, and we were linked by political and military pacts with some twenty countries outside of Latin America. It was the Korean War and not World War II that made us a world military-political power."[18] Or, from Truman's perspective, it was Joe Stalin who begat the US colossus.

Although Korea was called the ugly war that nobody won, the real victor in the conflict was Japan. Historian Yoneyuki Sugita writes that the Korean War came along like a "divine wind" for Japan and its devastated economy. "Japan received large orders from the United States to manufacture military supplies and to repair ships, tanks, jeeps, aircraft, and others . . . returning Japan to a prewar productivity by 1951 . . . and from a $300 million trade deficit in 1949 to a $40 million trade surplus by December of 1950."[19] Japan got acquainted with, and adapted well to, the war-imported assembly line and mass production techniques from the United States. Such an unprecedented opportunity to pull themselves out of an economic hell was not lost on the eager and industrious Japanese. They created their own divine wind. An unknown company named Tokyo Tsushin Kogyo, Ltd., known today as the SONY Corporation, acquired Bell Laboratories' transistor patent in 1954; and by July of 1956 Sigeru Takahashi and his mates at the Japanese government's Electrotechnical Laboratory had built Japan's first transistor computer, the ETL Mark 3—barely a year after Bell Labs' own TRADIC (TRAnsistorized DIgital Computer). Japan's electronics industry was on the rise and somewhere in there a little homage might be due Comrade Joe.

From 1950 to 1955 was a mere five years for 450-odd military bases to suddenly pop onto the face of the earth. Fast work, to say the least. And with those 450-odd bases came the need for expansive lines of military communications by radio, teletype, and telephone among far-flung bases and their headquarters, for airborne communication and navigation, for radar systems, and, yes, for computers . . . dependable computers that could do more than just count. SAGE-type networks would eventually find themselves watching European skies for NATO as well as Asian air space from a newly remilitarized Japan.

Another ugly but important war took place in America's courtrooms. The Justice Department's antitrust actions against AT&T and RCA in 1949, and then with IBM in 1952, each resulted in consent decrees in 1956 that blew up their high-tech patent monopolies. No one demobilized after this war. The big three bowed low to the court but then regrouped in subtler ways for a quick return to business as usual. Time was of the essence for the smaller and newer electronics wannabes. AT&T was allowed to stay only in the field of telecommunications and had to license its patents. As for RCA, "Trustbusters," reported *Time,* "charge that the company uses its pool of 10,000 radio-electronic patents to keep industry in the dark about new developments, and forces licensees to pay for RCA patents they do not need."[20]

IBM was dealt a huge blow. Its former practice of only leasing a machine, and never selling one, was struck down by the Southern District Court of New York. An IBM 700 Series computer that formerly leased for $30,000 a month could be bought for $1,700,000. Worse yet for Big Blue, it could not hold customers in bondage to service and parts available only through IBM. Services and parts could in many cases add an extra $4,000 a month to a computer lease. If the mean time to failure on an IBM 701 was 20 minutes, then parts and services could well be a fantastically lucrative business. The 1956 decree required IBM to train personnel and to provide technical manuals for anyone who owns, repairs, maintains, or distributes IBM electronic data processing equipment. And the decree also ordered IBM to grant unrestricted licenses to companies desiring to use its patents, and enjoined IBM from instituting any suits for patent infringement that occurred prior to the date of the decree.[21] The end result of the consent decrees was that they afforded the little guy a slim chance to get into the game before the door slammed shut again, which it did but not before they did.

Forrester's magnetic-core memory was swept into the IBM black hole of patent control. IBM would have been insane not to try to control in some manner, shape, or form Forrester's invention. Using magnetic-core memory in SAGE government computers was legal, but its use for commercial purposes was not, which IBM desperately needed to do for replacement of unreliable electrostatic memory. The company turned to its Intellectual Property department, which kept a worldwide patent watch for any infringer or potential patent menace to its stable of inventions.[22]

Run by an IBM loyalist of 25 years named James Birkenstock, the Intellectual Property department took considerable interest in the invention of its newest best friend from the Barta Building. Birkenstock and his department plus IBM intellectual property lawyers swept the world clean of any opposition, potential opposition, or even any seemingly insignificant hint of opposition to IBM's commercial use of magnetic-core memory. Armed with cash and clout and a smile, Birkenstock swept away potential threats—a pile of cash went to a Los Angeles public works inspector who invented out of his basement laboratory; An Wang got $500,000; An Wang's co-developer Way-Dong Woo was bought off; and a German named Gerhard Dirks got $1,000,000 plus IBM help emigrating to the United States with his wife, and a job in sunny California at IBM's San Jose plant.[23] IBM's own Munroe (Mike) Haynes attempted to put his core memory design into the mix, but it was scrapped by IBM as unworkable. And anyway, why waste the effort when Forrester's prize was there for the taking? The only possible threat to IBM's use of core memory came from RCA engineer Jan Rajchman, who had a magnetic-core memory patent pending at the US Patent Office. Since Birkenstock and IBM had a previous cross-license agreement in place with RCA, IBM was free to use the technology. Tom Watson Jr. ordered his company to discard all electrostatic memory for the 700 series and retrofit them with core memory.[24]

But what about Forrester and MIT? They were in line to get zero. IBM begged innocence with the limp excuse that Rajchman was first in line at the Patent Office (1950), Forrester second (1951), and that was that. The waiting was over in 1955 when Research Corporation, the Cambridge patent licensing agent for MIT, filed suit against IBM for "willful infringement of the Forrester patent"—supposedly without informing MIT it was doing so.[25] Tom Watson Jr., who sat on MIT's board, promptly resigned his MIT seat; James Killian, MIT's president, who sat on IBM's board promptly resigned his IBM seat in return. And the royalty that Research Corporation sought from Birkenstock and his crew of patent sweepers was considered by IBM as outrageously greedy.[26] Something had to be done; no one at IBM dared rock the boat. Watson and Killian settled it old-boy style: IBM paid MIT a lump sum of $13 million; MIT paid off Research Corporation, and then promptly fired them. IBM would go on to produce trillions of ferrite cores and use them to make millions of magnetic-core memory planes. Birkenstock secured a fortune for his company; and IBM would rake in profits hand over fist until 1970. Later, during litigation, Rajchman admitted to Forrester's prior discovery of magnetic-core memory and also conceded that Forrester had been first to develop working core memory and first to install and run it in a computer. However, Forrester had been late in filing his claim within the prescribed limits set by patent law. With all the paper published by the bright boys on everything and anything at their laboratory, somehow—incredibly—this "slight" paper-filing oversight had taken place. Somehow too, Research Corporation had been less than diligent in failing to notice the elephant in the room. How could they have missed it?

In 1956 Jay Forrester finally prevailed in court and toted home his long-overdue patent: no. 2,736,880. First the SAGE contract, then Whirlwind technology, and finally core memory; the bright boys were IBM's "divine wind." Between 1954 and 1956, IBM's gross income soared from $570 million to $892 million; it added 22,000 new employees, split its stock five-for-four and declared a two and a half percent stock dividend. Consent decree or no, IBM was virtually unstoppable. By 1956, IBM was no longer a tabulator company but "the world's largest and most profitable computer manufacturer, building the machines by the hundreds."[27] The same "divine wind" blew MIT's way as well: "Royalties to MIT from nongovernment sales amounted to $25 million, as MIT licensed the technology broadly."[28] In 1970 up popped a newly formed company calling itself the Intel Corporation. Intel released its 1103 transistor chip, the first DRAM (Dynamic Random Access Memory) chip, which deep down inside was quite akin to Forrester's core memory concept but made really, really, really small. By 1972 the 1103 was the best selling semiconductor memory chip in the world.

In the early spring of 1950, as Forrester worked with the Valley Committee and watched their faces daily growing more confident and enthusiastic about his mammoth computer, he grew less confident. He knew full well that the electrostatic tube shop in the Barta Building was an exercise in futility. But he had to keep up a good front and wait. As late as 18 months after his first introduction to the Valley Committee, August of 1951, his lab's snappy-looking *Whirlwind I* brochure dedicated a full page of praise for electrostatic storage.[29] Complete with a photo layout of tube makers in white shirts, ties, and surgical gloves handling tube assemblies with medical instruments, the brochure's glossy outward appearance belied the undercurrent of unease that preoccupied Forrester. He also knew full well that when he fitted Whirlwind with its first set of memory tubes, they were doomed to early burnout. Reliability meant only one thing: magnetic-core memory. Core memory existed only in the pages of his notebooks.

Everything that the bright boys had worked six years to achieve was riding on those notebooks, as was George Valley's reputation, as were the hopes of Muir Fairchild and Gordon Saville's air defense, as was the rapid growth of a young computer industry. How could so much be riding on one man's notebooks? A new decade was dawning. Joe Stalin was ready to play his role, as was the Justice Department. What about Forrester?

Lincoln Laboratory

George Valley's encounters were unusually interesting and seemed always to have surprising impact and consequence well beyond their intended purpose. One of his most conspicuous encounters was a December 15, 1950 luncheon with Louis Ridenour at the Chief of Staff's table at the Pentagon. Valley was in Washington on other business when Ridenour passed a note to him requesting a rendezvous. Valley and Ridenour were not strangers. Much to the contrary, they were friends; both were Rad Lab grads and members of von Karman's Air Force Scientific Advisory Board (SAB). Handsome, a meticulous dresser, athletically fit with a rugged edge, Ridenour looked none the part of the brilliant physicist that he was. A professor at the University of Illinois and former overall editor of the Radiation Laboratory series on which Valley had also labored, Ridenour was in 1950 the first chief scientist of the Air Force. It was as both friend and chief scientist that Ridenour asked Valley to lunch. He had a favor to ask, a proposal to make, and some advice to share.

Valley wasn't completely in the dark about Ridenour's purpose for the meeting. He was well aware of the traffic of interest afoot regarding an expansion of the Valley Committee. He had already chatted with some ex-Rad Lab veterans, who had signed on with Air Force Headquarters and were hoping to resurrect a semblance of the Rad Lab, especially to counter a rising Soviet threat.[30] The raging Asian war that by December was six months on and not going well was of particular importance. The Inchon landing had taken place the previous September and the North Koreans looked defeated. But, the Chinese suddenly entered the war and Russian pilots were flying MIGs. As the two men

Lincoln Laboratory (Building F blockhouse, center left), 1955. (Photo courtesy of MITRE Corporation.)

met for lunch, the First Marines were retreating from the frozen Chosin Reservoir to a port of escape at Hungnam 78 deadly miles to the south. Missing from the war was an OSRD-type organization, the kind Vannevar Bush had advocated in *Science: The Endless Frontier*, writing "our defense against aggression demands new knowledge. This essential new knowledge can be obtained only through basic scientific research."[31]

The Air Force and their chief scientist were trying to jump-start their new-knowledge machinery of basic research. A month after the war began, the Air Force, tipping its hat to Valley, mirrored the Valley Committee by starting a similar committee to review the guided-missile program.[32] At the time, Ridenour chaired a SAB committee putting together the new Air Research Development Command (ARDC) where there was strong interest in setting up a laboratory for basic research and development.[33] By May of 1951 ARDC would become a major Air Force command under Major General David Schlatter, and would be looking for a steady supply of modern electronics for command and control of air defense, SAC, tactical operations, and the brace of new air bases around the world. All of this was very much in keeping with von Karman's R&D master plan in *Science: The Key to Air Supremacy*, and *Where We Stand*; all of which brought George Valley quickly front and center into the eye of the movement. Valley and Ridenour met on a Friday, a week before Christmas, a good time to reflect on the remarkable accomplishments that Valley's team had rung up just short of one year since the Valley Committee began on January 20, 1950.

One of the moving parties for the new laboratory was Ivan Getting (then Air Staff Assistant for Development Planning), who admired the way that the Valley Committee cut to the chase when applying evolving technology to meet Air Force operational needs.[34] It was very Rad Lab-esque to Getting and others. The Valley Committee "was not inhibited by the bureaucratic inhibitions and restraints characteristic of military organizations."[35] As Valley put it, his old Rad Lab mates "demanded" a show-and-tell of the Valley Committee's air defense toys. They liked what they saw and came away convinced that a new Air Force-financed, air defense laboratory was necessary.[36]

Getting made his position clear to both Valley and Ridenour. In October, MIT's Stratton related to Killian a conversation with Ridenour in which Valley's work was spoken of as "our brightest hope in the field of air defense," and that there was a need of a laboratory to further that work.[37] A month before the Pentagon lunch, November 20, 1950, Ridenour sent a memorandum to General Saville requesting that he move on the air defense laboratory "by negotiating a research contract with a suitable institution in the Cambridge area . . . and MIT has indicated that they would consider taking such a contract."[38] The memorandum, titled "Proposed Augmentation of ADSEC Activities," called for a laboratory of 100 researchers and a budget of about $2 million a year. Valley had been copied on the memorandum. Lunch would be long, and Ridenour wanted to get a personal feel for Valley and his Committee's work. Successes were everywhere for Valley's Committee, so it was easy to give Ridenour the cook's tour. Ridenour already knew most of what Valley would tell him, but that is where the favor and the proposal came in. Ridenour was there to draw out Valley's commitment to an ongoing leadership role in air defense after the Valley Committee came to an end.

The success of the September radar tests was critically important. Charles Draper's Instrumentation Laboratory had a hangar at Hanscom Field with a pilot and a B-26 for use in the tests. Harrington and Forrester demonstrated that radar data could indeed be received from a phone line, manipulated by a digital computer, and then displayed on a cathode ray tube."[39] To Valley, the test was "proof of principle, and sufficient for ADSEC's purpose, which was to tell the Air Force what to do, not actually to do it for them."[40] What he had signed on for with Robertson during the taxi ride was now delivered, and as such, Valley could argue that his mission was at an end. Returning to cosmic ray studies and the good life was his for the asking. But he did not ask. And Ridenour was not about to offer it. Ridenour wanted more than proof of principle; he wanted an air defense system and he wanted Valley and his boys to make it happen.

Valley was willing to agree with Ridenour in large part because of the early flak such a notion had already stirred up at MIT. "Some decided to oppose the setting up of a new laboratory unless they could run it," then a "second group . . . began to attack ADSEC and me as incompetent."[41] "Jerome Wiesner and Jerold Zacharias," wrote Ivan Getting, "opposed the venture out of fear that it might dilute their own RLE [Research Laboratory of Electronics] programs."[42] Gordon Brown feared that RLE's work would atrophy;[43] and the Army and Navy "were fearful that their influence with MIT would be eroded."[44] When Ridenour asked Valley to draft a letter to MIT's president, James Killian, asking for just such an air defense laboratory, Valley was quick to oblige. "I completed it in about an hour, and Ridenour spent another fifteen minutes recasting it into appropriate general officer's diction."[45] Ridenour had it typed, and then got Vandenberg's signature on it by four o'clock that afternoon. The Valley-Ridenour-Vandenberg letter went that same day directly to Killian: "The Air Force feels that it is now time to implement the work of the part-time ADSEC group by setting up a laboratory which will devote itself intensively to air defense problems."[46] What had been merely a "notion" before lunch was now damn close to a command from the highest officer in the Air Force and one of the most powerful men in Washington, whose F-86 pilots were tangling daily with MIG-15s over Korea.

Air defense was big stuff, bigger than MIT and Killian. Senator Henry Cabot Lodge Jr. told his fellows on Capitol Hill what they already knew, that the air defense of America was "so feeble as to almost invite attack." Diatribe or not, MIT was not going to buck that current. Financially, morally, and academically MIT was left without a scintilla of choice. Before putting the university's official imprimatur on the Vandenberg request, MIT first needed a little face saving, to placate its upstaged military sponsors, and to cheer up unhappy researchers. MIT asked former Rad Lab second-in-command, F. Wheeler Loomis, to come in from the University of Illinois to head up a study group that would assess ADSEC's work and make recommendations. It all took shape in February of 1951.

Occupying the upper floors of the recently purchased Lever Brothers Building, in space scheduled for MIT's new School of Management, the well-respected Loomis kicked off the examination of the ADSEC extension into a permanent air defense laboratory. Project Charles—as it was called because the Lever Building overlooked the Charles River—consisted of 28 full-time scientists and engineers; 16 consultants; air officers from Great Britain, Canada, and the United States; 2 business managers; 18 secretaries; a librarian; a mechanical engineer; and two assistants. Among the group leaders selected for Project Charles were Wiesner, Zacharias, and Gordon Brown. Attending, as two of the consultants, were Ridenour and von Neumann. For six months Project Charles closely examined all findings. Valley wrote everything into its final report—in which there was not one proposal to improve upon the original ADSEC recommendations.[47]

On April 20, 1951, the bright boys tossed the clincher at Project Charles. With Forrester's first set of electrostatic tubes in Whirlwind, they performed the first-ever air detection, tracking, and interception—not once but twice more for good measure. Three times in one day. With the B-26 as the interceptor, the bright boys and Whirlwind tracked and scanned a T-6 trainer and a Beechcraft twin-engine C-45, each time maneuvering the B-26 to within 500 yards of the targets.[48] Whirlwind's detractors were totally overwhelmed by the air defense display. "Much of the credit goes personally to Jay Forrester," said Valley. "Each day, after enduring hours of Project Charles, he nursed his balky storage tubes late into the night."[49]

Project Charles reaffirmed ADSEC. All that remained was to divvy up the new prize and give it a name. "At MIT's insistence . . . the three services contributed to the budget in roughly the following proportions: Army, Navy, Air Force: 1, 1, 10. In this way, the Air Force was allotted most of the services of the new laboratory [and paid the most] . . . the opposition group of scientists were granted a dominant influence in the laboratory, because most of the directorships went to them, as well as control of the Army and Navy money."[50] Project Charles concluded warmly for everyone on Winter Place at Boston's Locke-Ober Restaurant over dinner and a celebratory toast of Napoleon brandy. At the nexus of Lexington, Bedford, and Lincoln, Massachusetts, a patch of land near Hanscom Field was selected as the site for the new air defense laboratory. Since MIT already had a Project Lexington and a Project Bedford, the freshest name left was Project Lincoln. As bulldozers ground into action and clawed away at Lexington's forest clearing sites for buildings, the name Lincoln Laboratory was selected to go above the front door. The bright boys had outgrown Cambridge and would now have a complex of buildings and their own idyllic chunk of suburbia.

Sons of Whirlwind

Project Lincoln got rolling in Cambridge in July of 1951; the official Air Force contract was signed in September. Buildings were promised to be ready in the Lexington woods by April of 1952. Awaiting the move and collocating together on campus were the elemental parts of the air defense laboratory: radio communications, long-range communications, radar, solid-state physics, airborne early warning, systems components, and ordnance.[51] One thing was for sure in the transition: whenever and wherever the bright boys moved, Whirlwind was not going along with them. Whirlwind was wonderful but, as with many inventions, it was a one-of-a-kind machine that was impossible to mass-produce. The boys needed a production prototype of Whirlwind that a factory could better duplicate and manufacture.

The bright boys immediately began to plan for such a machine, one they called Whirlwind II. Whirlwind II, however, would never happen. Project High came along in the fall of 1952, making IBM the provider of a reproducible Whirlwind. The new Whirlwind would come as the identical XD twins: XD-1 going east for the bright boys and XD-2 staying put in Poughkeepsie. When production of these prototypes bogged down Project Grind ground into being in June of 1953 to "unbog" Poughkeepsie. Forrester and Everett used the meetings to get directly to "the technical design problems, making it clear circuit by circuit and system by system" what IBM's engineers needed to do.[52] A delivery date for XD-1 was set for January 1, 1955; a building to house XD-1 was also planned for Lexington, which would be named Building F, with an anticipated completion date also slated for

January of 1955. Still another name change took place when XD-1 took on its official military nomenclature: AN/FSQ-7. The new name proved to be a mouthful, so it was nicknamed the FSQ-7, then shortened again to just the Q-7. Since the new machine would not be around until 1955, old faithful in the Barta Building would pull the load until then. Project Lincoln began moving into its new street address at Lincoln Laboratory beginning in the spring of 1952, continuing into 1953, and completing in 1954. Left behind in the wake, the bright boys began readying for their biggest test yet, the famous Cape Cod System.

Another smaller Whirlwind, called Whirlwind 1½ or sometimes called Whirlwind 1A, would go into development on nearby Vassar Street in MIT's Whitemore Building, formerly the Whitemore Shoe Polish Company. Ultimately, this machine also underwent a name change, becoming known as the Memory Test Computer or MTC. Ken Olsen, a former graduate student of Forrester's, who like William Papian wrote his master's thesis on magnetic memory, headed the MTC design team. Assisting on the MTC design was a Forrester new hire (1951) and newly anointed bright boy, Wesley Clark, fresh in from the West Coast where he'd been working at Hanford on atomic reactor research. Clark soon proved to be another of Forrester's more prescient employee placements: he and Olsen would later migrate MTC technology to Lincoln Laboratory, where they would marry it all to transistors and a bit of new machine design to produce two remarkable computers, the TX-0 and TX-2.

"It was supposed to be an honest-to-goodness computer that would really run and test the memory, but not a computer that was designed to be useful," remembered Olsen about his days constructing the MTC. "I was given the job of building the computer just as soon as my thesis was done. I think I was still a graduate student and it cost a million dollars. I can remember being impressed how much a million dollars was. How much work it took to spend a million dollars."[53] If and when Forrester got his magnetic-core memory together, the MTC would be used to test how well it worked or didn't work. If all went according to plan, electrostatic storage would be removed from Whirlwind and replaced with core memory. As Olsen remembers, within one day of the MTC cranking into action in May of 1953, the electrostatic storage tube facility was shut down and everything switched to core memory.[54] Bright boy Pat Youtz, who had led the tube laboratory, immediately switched over to leading the effort for Whirlwind's many display screens in the Barta Building's top-secret Room 222.

Even with the MTC's success, Olsen was not done with the machine: he and Clark got the chance to build it again—this time with transistors—in an isolated basement in Lincoln Laboratory.[55] Fueling their ability to evolve the Memory Test Computer through to the newly emerging world of transistors was the insightful creation of Group 63, the Advanced Development Group at Lincoln Laboratory. Instead of cloistering all research in Building F as military, and therefore classified, the bright boys reached out to innovation once again. "When the Digital Computer Laboratory was formally made Lincoln Lab's Division 6 for classified work under the SAGE umbrella," remembers Wes Clark, "Forrester and Everett saw to it that basic, unclassified research and development of computers would not simply disappear in the process." To that end, they organized Group 63. Initially under David Brown's leadership, and later Bill Papian, Group 63 brought together those who had worked on the MTC plus others doing research and development on memory and circuits. There were even subgroups with physicists, chemists, and other engineers as team leads.

There was also a practical side to the creation of Group 63: survival. The concern was "that SAGE was consuming Division 6," recalls Bob Everett, "and if we did not do something we would end up with no future work at all. We, therefore, set aside a selected group of staff with the mission of continuing basic work on computer technology and protected it from the demands of SAGE. SAGE provided the money." With SAGE as paymaster, as many as thirty staff then had the chance to forge ahead with the likes of "transistor machines, core memories, time sharing, computer games, printers, and many other things," safeguarding a future in basic research while keeping apace or even keeping ahead of ever-newer breakthroughs in technology—all of which was beneficial to both SAGE and Group 63. Much of their work would inure to the paymaster as payback, as with "the 65,000-word memory that was of great value to SAGE."

As things would eventually turn out, this insightful creation would prove to be a wonder box of unintended consequences. With Clark handling the logic design and Olsen the transistor circuits and hardware, they proposed a new machine that they tabbed the TX-1. "The idea was soundly rejected by Jay and Bob Everett," says Clark, "on the grounds that we didn't yet know enough for such a big step." Back to the drawing boards they went, this time proposing what Clark refers to as a "primitively simple computer." This new-age MTC would go by the name of TX-0 (the TX referring to Transistor eXperimental, and pronounced as "Tixo"). Olsen, naturally influenced by Whirlwind and the MTC's packaging, circuits, and toggle switches, rebuilt MTC's circuitry using 3,600 of Philco's surface barrier transistors at $80 a pop ($288,000).

The TX-0 turned out to be a 5-MHz, parallel, general-purpose, stored-program, digital machine. It had a cycle time of six microseconds and was capable of performing better than 80,000 additions per second. One

TX-0 transistorized computer at Lincoln Labs, 1956. (Reprinted with permission of MIT Lincoln Laboratory, Lexington, Massachusetts.)

objective in building the machine was to test and evaluate the use of transistors as the logical elements of a high-speed computer. The second purpose was to provide means for testing a large capacity, 65,000-word magnetic-core memory.

The TX-0 (1956) was a success that spawned in 1958 an even larger, more advanced descendant in the TX-2 (22,000 transistors). Wes Clark and his team designed the TX-2 for Ken Olsen's team to build. It was this TX-2—with Larry Roberts at the helm on the East Coast, networking westward to Tom Marill in Santa Monica at the helm of another Whirlwind successor, the AN/FSQ-32 (1965)—that was the original transcontinental hookup that heralded the beginning of ARPANET. "This demonstrated that a packet-switching network could work across the country, linking two different computers running two different operating systems, thereby making a key step in the evolution of what became the Internet."[57]

The TX-0 would serve Ken Olsen again. Impatient for the TX-2's completion, which wouldn't happen until 1958, Olsen left Lincoln Laboratory in 1957 and founded the Digital Equipment Corporation (DEC). His early line of PDP computers were reported to be advanced versions of the TX-0 wearing DEC logos.[58] Digital, as Olsen's company was known, eventually became the near equal of Snow White. These sons of Whirlwind—MTC, XD-1, AN/FSQ-7, AN/FSQ-32, TX-0, and TX-2—moved seemingly in every direction at once: changing their names, rearranging their form, physically moved, and re-purposed, yet still influencing everything in their paths.

Meeting notes from Group 63's Steering Committee from June of 1958 display the wide diversity of intended research: "data processing, pattern recognition, voice recognition and other self organizing system studies," which even included a stint with Charles Molner, of the Communications Biophysics Laboratory under Professor Walter Rosenblith, "to aid in the analysis of brain-wave data gathered from the auditory cortex of a cat's brain."

However, the true sons—and daughters—of Whirlwind weren't metal, glass, magnetic cores, or even transistors, but very human. Overjoyed that

Whirlwind was staying put in Cambridge were the hundreds who journeyed to the oracle for help, got it, and went away true believers. This
diaspora of the "converted" created a large, ever growing, and enormously
influential fan club that proselytized the great machine far and wide. In
the 1950s, if anyone wanted to know which way to the future, all one had
to do was to follow graduate students and others heading over to the ever-
popular 211 Massachusetts Avenue. Across the MIT campus and beyond,
researchers from academia, business, and industry lined up to get their
calculations done. If a problem was either too time consuming to do by
hand or totally intractable except by machine, Whirlwind would step in to
crunch the numbers. With plenty of calculations in need of such treatment,
Whirlwind was not only in demand but, as would be expected, became
an admired member of everyone's family. Although working on top-secret
government air defense projects, the bright boys offered machine time on
Whirlwind to anyone. Even "senior mathematicians had lost interest in
the application of analog computers for analyzing random processes. They
were switching to digital techniques,"[59] and although "Whirlwind was over-
crowded, almost no one wanted to use the much slower and less accurate
Differential Analyzer, so it was scrapped."[60]

The bright boys were putting the right tool into the right hands, and in
doing so, created a digital movement. An MIT grad student, Eldon Hall, and
his wife were two of many. Hall, who years later would design the Polaris
missile guidance system and the Apollo Guidance Computer that landed
astronauts on the Moon, got his first look at a digital computer in 1952
watching Whirlwind compute the antenna pattern for a dipole antenna
and display the results on an oscilloscope. "I learned that this monstrous
marvel was available free of charge to researchers with complicated problems."[61] His wife, needing to compute the eigenvalues of matrices, asked
the great oracle for help. "My wife submitted the elements of a matrix to an
individual at the Barta Building. He reappeared a few days later with the
results, truly a marvel when compared with the hours of labor using a mechanical calculator."[62] Hall was hooked. Later, when the Air Force required
the extreme accuracy of a digital computer for intercontinental missiles,
and Hall's lab "had no experience in digital control systems," Hall began
building them.[63]

For 22-year-old Enders Robinson, the allure of Whirlwind was extraordinarily fascinating and would lead him into a new career, one where
he would revolutionize the field of geophysical exploration—basically,
the search for oil. In 1952, as an assistant to MIT mathematician G. P.
Wadsworth, Robinson was given eight seismic records from Magnolia Oil,
the research arm of Mobil Oil, and asked to submit them to some, any kind,

of mathematical scrutiny. Just what sort of scrutiny no one had the foggiest idea. "The first time I ever saw a seismogram," he said years later in an interview, "it was not a pleasant feeling to look at that jumble of lines crisscrossing all over the place and know that you're going to have to try to find a pattern in the mess. In those early days, I had no idea about how oil was found and it wasn't an easy thing to pick up in Boston. I went to the library but at that time there were only three books on geophysical exploration. After reading them I felt worse off because I realized how little I knew."[64]

Like so many others searching for answers to tough questions, Robinson traipsed on over to the oracle in Forrester's Digital Computer Laboratory. Robinson's basic plan was to take the seismic records and apply to them mathematical methods based on Norbert Wiener's concepts for deconvolution and time-series analysis, which came from Wiener's recently published book from 1949, *Extrapolation, Interpolation, and Smoothing of Stationary Time Series*. Deconvolution is the algorithm-based process used in the techniques of signal processing and image processing. "In the spring of 1952, I went to Whirlwind with Howard Briscoe and put deconvolution on Whirlwind." And with that first brief encounter, the great machine quickly added yet another new convert to its ranks. "Whirlwind was good for signal processing because it could handle a lot of data . . . for signal processing, for deconvolution . . . it was very fast and efficient . . . it could handle a lot of data that the other machines of the early 1950s could not."[65]

For the young mathematician, the encounter was nothing short of life changing. He had hit upon seismic research's core concept: converting continuous seismic traces into digital form and then using mathematical methods to enhance the data. Soon after, embarked on a new career as a geoscientist, he next founded the Geophysical Analysis Group at MIT, which a decade later led to a digital revolution in geophysics. Today, his book, *Digital Imaging and Deconvolution: The ABCs of Seismic Exploration and Processing*, co-authored with long-time colleague, Sven Treitel, is still a must-have resource for every geoscientist.

These advances in digital imaging and deconvolution eventually found ways into yet other areas of observational science, such as oceanography, meteorology, planetary science, astronomy, and medicine—in fact, medical ultrasonic imaging is the direct counterpart of seismic imaging.[66] Although Projects Charles and Lincoln were soon presented the looming reality of trucking everything off to the suburban retreat that would become Lincoln Laboratory, the great machine was staying put, and Forrester and Everett made certain that research applications using Whirlwind's computing power were never abandoned at 211 Massachusetts Avenue. To the contrary, the pace in the Barta Building was near furious. In the four months between December 1952 and the end of March 1953, a total of 781 programs (exclusive of military applications) were run on Whirlwind.[67] Nearly 200 a month, which was even more remarkable given that military use of Whirlwind was a priority. "The military took about eight hours a day," recalled Robinson, "and since it was a vacuum tube machine, the maintenance took about eight hours a day, and the academic people were supposed to get eight hours a day, but it never worked out that way because sometimes the maintenance took more time."[68]

Put in charge of squeezing non-military applications into these eight hours or less was the energetic and ever resourceful bright boy, Charles Adams. In 1953, the 28-year-old Adams became one of the first systems programmers in the Digital Computer Laboratory's Mathematics group. Adams was particularly concerned with preparing routines for Whirlwind, including the assembly language program, the design of floating-point arithmetic, and utility routines for debugging. But it was as the head of the Scientific Engineering Applications group that he provided invaluable assistance to those eagerly awaiting a turn at solving unclassified problems in physics, chemistry, or engineering, or those from most any other discipline with the yen to take a dive into Whirlwind's world.

The gamut of problems that Whirlwind was asked to unravel would include not only those as in Robinson's deconvolutions but also excursions into areas like deuteron binding energy and wave functions for MIT's physics department; or transient aerodynamic heating of flat plates for the aeroelastic and structures laboratory; or optical properties of thin metal films for the chemistry department; or even writing a graphics program together with the visiting Maurice Wilkes, creator of EDSAC. Whirlwind was a magnet for attracting those searching for answers to what mathematicians call "hard" problems: those intractable problems that throw up roadblocks to progress somewhere, sometime, in most every discipline. And Adams was the master guide, busily leading one group or individual after another through Whirlwind, helping each to open doors of perception into worlds previously thought inaccessible. And, most often, succeeding.

For those a bit uncomfortable, unfamiliar, or downright skittish around the new field of electronic, digital computing, Adams was also there to teach the tricks of the trade to any (including military personnel) who wished to know. He taught a monthly programming course in MIT's electrical engineering department using the Maurice Wilkes, D. J. Wheeler, and Stanley Gill book on programming the EDSAC, as well as a course on advanced programming techniques. In the summer he ran an intensive, two-week program on computers and computer applications that included

Barta Building, Room 222, circa 1954.
(Photo courtesy of MITRE Corporation.)

an introduction to digital-computer coding, a survey of existing computers, applications, numerical methods, and advanced programming techniques, all supplemented by group discussions and by demonstrations and practice on Whirlwind.[69]

Ever the stylish promoters of the new digital movement, Forrester's Digital Computer Laboratory not only put out flashy Whirlwind brochures but also made a Whirlwind movie in 1953 titled *Making Electrons Count.* Part of the movie actually depicts how the typical supplicant approached the oracle. A physicist from the Retina Foundation comes with optical design formulas that Whirlwind successfully crunches for him. The film's voice-over narration then brags: "We have told you the story of one problem recently solved on one digital computer . . . you have an idea of the current importance of the digital computer as a new tool to help scale some of the hitherto insurmountable peaks which span the domain of man's activities."[70] "Hitherto insurmountable peaks" may have been a little over the top, even for 1953, but that's exactly what pilgrims sought and got from Whirlwind.

For one mathematician who made a pilgrimage to the oracle, just dropping off a problem and then returning the next day for an answer was not enough. Douglas Ross sought after a hand's on experience with man and machine working out the problem together. At the time, such thinking was a bit fanciful. Most people were thrilled to death at handing over a stack of IBM cards for an answer the next day. "The next day you got your answer back," remembered Olsen, "and it usually was that you'd made a mistake," which meant starting all over again with IBM cards. Ross, who came to MIT in 1951 as a freshman calculus instructor, saw Whirlwind's consoles and display scopes for the first time in the summer of 1952. "Significant events," he later recalled, "always are the rearrangement into new forms of otherwise ordinary and insignificant, routine happenings."[71] The Whirlwind viewing screens that were used for radar tracking and plotting coordinates of aircraft, Ross saw as personal workstations for interacting with a computer to solve problems or to accomplish a task. If something was wrong, then you and the computer worked it out right then and there. What J. C. R. Licklider contemplated in 1960 in his "Man-Computer Symbiosis," and what Doug Engelbart also in the 1960s saw as "human augmentation," is what Ross thought about while standing in front of Whirlwind in 1952 . . . nearly a decade ahead of Licklider and Engelbart. The way Engelbart articulated it was what went through Ross' mind that summer: "It's not just about putting a person in front of a computer. Computers are just a central part of the tool system. It's the combination of a tool system and human systems working in concert which is really powerful."[72]

Bright boy Jack Arnow introduced the young mathematician to the wonders of programming Whirlwind, and Ross quickly took it from there. By 1953, he had gained access to Whirlwind's secret room: "the locked, green double doors of Room 222," with all the air defense consoles, especially the E31 Console, the master control station that, if necessary, could intervene and run the entire machine. The all-black room with red light from indirect ceiling lamps was eerily fascinating for him. So too was the large wall image of an outline map showing the New England coastline; it was glowing at him from a screen and light canon of what was called the Area Discriminator. He wrote new code, devising programs to harness the huge machine to interact with him at a console. His purpose in Room 222 was to use Whirlwind to evaluate the performance of airborne fire-control systems, specifically, servo-controlled tail turrets in bombers. His fascination led him to use his experiences in Room 222 to build the first interactive, personal workstation in 1957 at Eglin AFB, Florida. During one typical adventure in Room 222 in 1954, Ross recalls the intercom crackling on and a voice booming out asking what he was doing alone in the darkened room. Ross had written a 200-line program to find and track a moving shadow. Hunched in a corner with his index finger pressed against a glass oscilloscope screen, Ross, his fingertip a blue-white spot of light, was writing his name "Doug," which was also writing itself on a screen in the Test Control room down the hallway.[73] It was the first-ever, hand-drawn input into a computer.

Cheerios

From February of 1950 to May of 1953 Forrester had his hands full of memory. He had bottles of it that he did not want but had to use, and cores of it that he could not use but wanted to. He had been filling up page upon page in his notebooks on magnetic memory since 1947; he had seen An Wang's method in 1949; he had submitted a paper on magnetic cores to the *Journal of Applied Physics* in June of 1950; and he had applied for a patent on "Multi-coordinate Digital Storage" on May 11, 1951.[74] The same day that he was filling out patent papers, the best he could do for his beloved machine was 16 bottles of electrostatic storage.

Forrester, with a bunch of Air Force money in his pocket, had Ernst Albers-Schönberg cooking up chemicals with the hope of finding the perfect core formula, and he backed that up with experimental core research at Arthur von Hippel's laboratory at MIT. He had neither bottles nor cores ready for Harrington's first capture and transmission of pings from the B-26. He had to dance with only one set of bottles in Whirlwind for the famous Project Charles demonstration. In 1952, he ordered the building of the Memory Test Computer to test memory cores that might not be available to test. He watched as Project Lincoln went from forest to concrete buildings, an entire facility waiting on a computer needing core memory for a national air defense system. He heard the gossip from "analogers," as Valley called them, who jeered from the sidelines hooting for failure.[75] As 1953 rolled into view, he had another major worry to add to a growing tale of woe: time was growing near for the anticipated Cape Cod System, when Whirlwind would be called upon to perform in an experimental air defense umbrella over a large chunk of New England. If he failed to deliver the goods, all the Air Force money in the world wouldn't be enough to cover his tracks out of town. And if his own spirit was not dispirited enough from all the growing concern, he still had to keep everyone else motivated with uplifting talk of an impending breakthrough.

The work of Albers-Schönberg was critical to Forrester's success. The ceramics maker hit upon the right ingredients and proportions of ferric oxide, ferrous oxide, and manganese dioxide, which he shaped in a press, fired, and then cooled to form the cores. The "old German ceramicist, who by guess or by intuition, or years of experience, mixed up his first mixture and hit exactly the same spot as all the research done at MIT."[76] This son of Paracelsus, the Harz, and the tradition and magic of searing crucibles worked his ferromagnetic ceramic cores ever smaller, using molds for better uniformity. Some called them "magnetic torroids;" "Cheerios" is what Olsen aptly called them because they resembled the size and shape of the breakfast cereal.[77]

The most important step came next—wiring them into an array, like a miniature chain-link fence with Cheerios strung throughout the links. "The core memory idea was not the idea of storing information in a core. That had been done before," Olsen said in an interview years later. "The clever idea of Jay Forrester's invention was the way of selecting the core."[78] Forrester created an array of cores that were 16 by 16 equaling 256 cores, and through the center holes of each were threaded four sets of wires: one set of wires ran north to south, another set east to west, with a third and fourth set running diagonally through the holes. Applying a selective current to any select wire made it "possible to choose any core within the solid array and to determine its existing state of magnetization or to establish it in either of its two desired states."[79] So magnetized in either of two opposite directions (clockwise or counterclockwise), each core could then be used for storing a bit of information. This "coincident current memory" made it possible to address any specific core for reading or writing information into that core. The two "desired states" were the equivalent of being "On" or "Off" or, in binary, a "0" or "1". "The ability to pinpoint specific intersections or addresses within the core rings, from which information could be stored and then recalled at random, created an unparalleled innovation in computing. The computer's central processing unit and its memory of stored data, procedures and programs, could now be operated interactively. Random access memory was born."[80]

The Memory Test Computer first tested the 16 by 16 core plane. It worked. The next test would be on a 32 by 32 core arrangement for 1,024 cores. Another success. From April into June of 1953, a 32 by 32 core arrangement stacked 17 planes high—the anticipated memory size for installation in Whirlwind—worked to perfection.[81] On July 1, 1953, bright boy David Brown's group ordered 250,000 cores to be sent to IBM for the core memory fabrication and installation on the XD-1 and XD-2. For Whirlwind, its first bank of magnetic-core memory was wired in on August 8, 1953, the second on September 5, 1953 . . . just weeks previous to running the Cape Cod System's first tests. Forrester had come through in a big way.

Guys like Doug Ross were ecstatic because memory now cycled at 9 microseconds versus 23 microseconds for electrostatic memory and accomplished 40,000 operations per second instead of 20,000.[82] "The memory never failed," remembers Forrester, which was a decided bright spot in an otherwise fragile new technology, "and Whirlwind had marginal checking to find drifting components before they caused trouble."[83] The bright boys, Project Lincoln, and the Air Force were ecstatic; IBM and the rest of the computer industry were jumping for joy as well, because they could then trash all the damn bottles forever. It was the breakthrough that the industry needed in order to move digital computing into complete primacy.

Valley's respect for Forrester and the bright boys was sky high. Magnetic-core memory, he said, "was a classic story of luck and pluck: a true epic out of the nineteenth century. The cores developed while I watched, and like a boy reading a Horatio Alger novel, I was inspired."[84] Of course, Valley did more than watch: he got the money—millions, he said—from the Air Force. The Air Force bet on a very long shot, and won. "Magnetic cores were simply not favored by the smart money," said Valley, "it was unlikely that perfecting them would have been regarded as commercially profitable."[85] Philips Laboratories, the vaunted Dutch high-tech giant and world leader in ferrite technology, called the idea of core memory an "American exaggeration." Even Bell Laboratories had shunned magnetic materials. Imagine the collective gulp they all took when Olsen cranked up the MTC and it purred, and when the bright boys trucked over 250,000 cores to IBM?

As great a year of discovery as 1953 was for Forrester, it was also a harrowing year where everything seemed time compressed, filled to the brim with breakneck activity, deadlines, tension, and heavy responsibility. The false attack at Nunivak and the country's first-ever nationwide Air Defense alert happened the previous April. The fallout in 1953 was that the guy who pulled the alarm, Air Defense commander General Benjamin Chidlaw, was over at the University of Michigan's Willow Run Facility ogling their air defense system. Whirlwind's competition was not much but it had the attention of the powerful. And sometimes that's all an inferior idea needs in order to topple a great one. Willow Run caused a ripple of dissent to move through the Air Force, eddying up uncomfortable fears in both Valley and Forrester.

Other airmen, majors and colonels, who worked with and firmly believed in Whirlwind, sat Valley down for a talking-to late in 1952. The Barta Building was abuzz with important air defense activity, much of which was particularly confusing for these officers in charge. Between August 11th and August 30th, Whirlwind was completely shut down and refitted with all the elaborate Input/Output hardware needed for the air defense system: viewing screens, operator consoles, Charactrons, light guns, magnetic drums, modems, telephone and radio communications, and all the other new accoutrements of tracking, scanning, and interception.[86] The officers wanted a blueprint, a document that totally visualized for them what this air defense Direction Center would look like, so that they could more fully envision the military aspects of their role in it all.[87]

Valley could see the reasonableness and sincerity of their request, one which he and his engineers had totally overlooked. Valley admitted that he and his engineers had assumed too much of the officers "treating them all like General Jimmy Doolittle, who has a doctor's degree in aerodynamics."[88] Like the layout for a new home, the buyers wanted to see the architect's drawings and plans, and not make do with verbal descriptions and grand talk. Grand talk certainly would not do when reporting to the high command who would want to hear every detail—all the specifics, especially where the general's command chair would be. Technical Memorandum 20 or just TM-20 was the result. When complete, they asked the recently retired Gordon Saville to critique their handiwork. As Bob Wieser remembers it, Saville, all of "five and a half feet tall, feisty . . . and with a strong voice," strode to the head of a meeting room table and slammed TM-20 down. "You are," he bellowed, "the worst damn salesmen I ever met. This report is *stinko profundo*." Wieser recalls listening to Saville very carefully, and began "to understand that it's one thing to explain something that lies outside a person's experience and yet another thing to explain something that lies outside a person's imagination."[89] They rewrote TM-20 . . . and Saville smiled his approval.

As necessary as TM-20 was for the Air Force officers, and as huge as was the impact it eventually had on the future training needs of the air defense system, Forrester needed, once again, to yank himself away from his pursuit of core memory and put together a how-to manual. Officially it was titled, "A Proposal for Air Defense System Evolution;" unofficially it was a proposal for knitting together Air Force officers and engineers into a team. Valley, Boehmer, Harrington, Forrester, Everett, and Wieser wrote it and delivered it in January 1953. For Forrester it was just another distraction, albeit an important one, but nonetheless a time-consuming distraction in a year filled with distractions. The effort was worthwhile. TM-20 was widely distributed in the Air Force and made an impact.

In March, Generals Partridge and Putt, the heavyweights of the Air Research Development Command, visited both Willow Run and the Barta Building, and on May 6th declared Whirlwind the champ.[90] Then, to add even more hurry-up in Forrester's rush to make Whirlwind ready, the Soviets added their own touch of anxiety. On August 12, 1953, they lit off the big one, a hydrogen bomb, the H-bomb. Equaling one million tons of dynamite, one of these big bombs dropped on, say, St. Louis, would have the destructive force of 70 Hiroshima-type atom bombs going

South Truro, Massachusetts radar station, circa 1958.
(Photo courtesy of MITRE Corporation.)

off simultaneously, with a flash brighter than the Sun, and with a thermal flash of over 1,400 degrees Fahrenheit (800 Celsius) traveling at the speed of light. The folks who got real jittery during the Nunivak scare, including the general population, suddenly saw not just mushroom clouds but apocalypse. The bright boys were center stage. Air defense was a must!

The bright boy most pleased with the success of Forrester's Cheerios was Bob Wieser, who was in charge of the Cape Cod Direction Center in the Barta Building. His big day on stage was to present to the Air Force a functionally complete experimental prototype of how an actual SAGE Direction Center would work. In a world just recently introduced to digital computers, where human interaction with a computer was exclusively done by loading IBM cards into the machine and then waiting for an answer, Wieser would perform by conducting an orchestra of 50 manned workstations and monitors (what were then called consoles). A cabinetmaking shop across the parking lot from the Barta Building built the consoles, and then Wieser and company crammed them full of electronics. Having two of Forrester's 1,024-core memory banks installed in Whirlwind seemed a godsend. The operator consoles had to be integrated by software, which, he said, was "the largest real-time control program ever coded . . . coding in machine language, since higher-order languages had not yet been developed."[91] Having solid memory to rely on was vital when simultaneously integrating hardware, software coding and debugging, and training Air Force personnel on the use of the equipment. When the September tests rolled around, Wieser was ready.

A long-range FPS-3 radar was installed near the tip of Cape Cod on a hill in the town of Truro, Massachusetts; gap-filler radars equipped with Harrington's "slowed-down video" went up to the south and north of Boston at Scituate and Rockport, Massachusetts. Each site also had an IFF (Identification Friend or Foe) system multiplexed to the radar information. Large spools of telephone cable dotted Route 6 leading to Truro, as well as the roadways leading to the other radar site—spools that would be unwound to connect up all of the far-flung sites. The bright boys' modem, the Harrington and Rosen 1,800 bits-a-second modem that AT&T scoffed at as impossible, was a reality. "They said if one drives a telephone line over 600 bits per second, one has arrived at the end of the flat earth," recalled Rosen.[92] Soon after, "the Bell Labs people adopted my modem" as the company standard. Networked together with telephone lines and modems, the system was in readiness. All of the sites' data traveled over dedicated telephone circuits to the Barta Building, right into Wieser's console-crammed Direction Center. This setup was, in miniature, a test-bed for mock air attacks on Boston, a major US city with a large population. If the system played well at protecting Boston, the model could be then extended out to cover all of North America. For the Air Force, this command and control of air power using a computer and interactive consoles was the wave of the future. Amid the glowing confines of Room 222 with its Charactron screens, Area Discriminator, and indicator panels of dancing lights, the Air Force could see the air operations center of the future.

Wieser's room performed wonderfully well, and continued the performance as the radar network expanded south to New York at Montauk, northwest to Derry, New Hampshire, and as far north as Brunswick, Maine. SAC B-47s attacked Boston, while Air Force F-89s and F-86s, and Navy F-3s successfully intercepted each bomber in each formation. By 1954, with the inclusion of automatic ground-to-air data links, Whirlwind was guiding interceptors to targets using the aircraft's autopilot, which was a prelude for future interceptions using pilotless BOMARC missiles. As Wieser said of his mates: "The workers were young, bright, enthusiastic, and very much aware that they were working on the leading edge of something new and important; they were learning on-the-job skills that schools did not teach. The hours were long, the camaraderie was close, and everyone wanted to make it work."[93]

For Gordon Saville, this was the culmination of his entire career blooming out of an old brick building in Cambridge. Somewhere, Forrester must have been reflecting back a bit on the long journey that had begun back in 1947 with L-1 and L-2, when he and Everett announced that such a world would come to pass. Back then, people brushed them off with a courteous smile. Not any longer. Herman Melville wrote, "It is better to fail in originality than to succeed in imitation." Maybe. But oh, how sweet it is to succeed in originality.

Chapter Nine
Voices in the Machine

*There are a great many things that all of us
could do much better if we could only apply
what the wisest of us knows.*

—Edmund Berkeley, 1949

Giant Brains, or Machines That Think

The XD-1 arrived at Lincoln Laboratory on January 5, 1955 and was assembled and running by January 12th. Building F, a windowless, concrete, three-story blockhouse attached to Lincoln Laboratory, would be its home for the foreseeable future. Building F was the model Direction Center for all that would follow it, complete with rooms for radar mapping, aircraft identification, weapons direction, battle station, and command post. XD-1, the new cookie-cutout Whirlwind, all spiffed up with 64k of magnetic memory courtesy of IBM, and sleek looking, like a new Pontiac fresh off the assembly line, tucked its generous proportions easily into the cavernous citadel.

Somehow, during its transformation in Poughkeepsie from one-of-a-kind original to assembly-line knockoff, the machine had lost some of its charm. Switching on the XD-1 was a bit anticlimactic, more like switching off an era. The bright boys respected and admired the new creation, but there was something endearing about the old machine that the new one could not match. "I felt that the pioneering days in computers were over,"[1] recalled Forrester in an interview 40 years later. Building the monster in the Barta Building was where the real fun had been. Things would never be quite the same again. From here on in, it would be perfecting the assembly of factory-made Q-7s for delivery, and fulfilling the contract to

Ret. Adm. Edward L. Cochrane (MIT's vice president for industrial and government relations), George Valley, Maj. Gen. Raymond C. Maude, and Col. Dorr Newton at press conference announcing the SAGE system for continental air defense. (Photo courtesy of MITRE Corporation.)

network the Direction Centers across the continent. The real action, the zest for experimentation and discovery, was over when the last of the Cape Cod System tests concluded.

Forrester would be gone in a year. Valley left the very next year to be chief scientist for the Air Force. Everett checked out a year later. The following year, January of 1959, the bright boys would split up forever. Some remained behind at Lincoln Laboratory to continue on in high-tech electronics, some left to seek new horizons elsewhere, and still others migrated a few miles away to join Everett at the newly formed MITRE Corporation to work the air defense contract for the Air Force. The partings were mixed. Many times as a technology matures and moves into wider acceptance, its originators and developers get chilled out of the flow of things. They lose control. They then move on or are moved out. Also, too long their own bosses, the aspect of heeling to new management and administrative masters was something some would never abide.

Plain to see for most was the fact that private enterprise and not universities was the future incubator of new computing machines. "I felt I had been in the field long enough," reasoned Forrester. Indeed, ten years of building Whirlwind and managing the bright boys, plus the war years in the Servomechanism's Lab, was a long time. "I think you can argue that the ratio by which computers improved in the decade 1946 to 1956 was probably bigger than in any decade since."[2] Mulling over an offer from MIT's Sloan School of Management and nurturing a new direction for his talents, Forrester retired from the computer wars in June of 1956.

Wildcats on their own for a very long time in Cambridge, it was inevitable and understandable that the new order of things would affect them. In 1954, *Time* reported on a survey conducted for *Fortune* magazine by Francis Bello that sought to find out "What kind of a man becomes an outstanding scientist." Far and away the most common "characteristic of outstanding young scientists [under 40]," he reported, "is a fierce independence. This is invariably coupled with a strong desire to work on the most crucial problems in their field."[3] For a long time the bright boys drank heartily from both of those wells. But, as they said in their own brochure: "Digital computation is a new and growing field and a young and rapidly developing profession." Certainly there must be other wells equally as sweet and heady as Whirlwind in that bright new world of digital computing out yonder. One never knows. The odds seemed in their favor and many went hunting for something few would find again. As Wieser mused years later: "Sometimes I ask myself why this was such an interesting experience, the like of which I haven't had since."[4]

The experience, much muted, continued on for the bright boys at Lincoln Laboratory. They were needed. They stayed a few years to finish what they had started. There was after all a national crisis at hand that demanded an air defense umbrella over North America. Certainly their patron, the Air Force, was watching in earnest, waiting for a solid return on its investment. The bright boys could also take pride in the fact that they were now founding fathers, ascending from cocky wise guys to venerable pillars of digital computing. The acres of cinderblock walls and miles of black and white floor tiles that was the new laboratory were there because of them. If Yankee Stadium was affectionately known as the house that Ruth built, then Lincoln Laboratory was the house that Whirlwind built. Without Whirlwind the Lexington woods would still be woods. And without the bright boys there would certainly be no Whirlwind. Even the name Whirlwind was quickly receding from common parlance. In 1954, what formerly was Whirlwind, then the Cape Cod System or Lincoln Transition System, donned its official name: Semi-Automatic Ground Environment or SAGE. At a press conference at Lincoln on January 16, 1956, SAGE was formally introduced to the public. And from that day forward, the machine that powered it all would be referred to as the SAGE computer.

With SAGE in the house Lincoln grew rapidly. In terms of revenue it soon dwarfed its parent; it's 2,000 staff, of which 700 were science or engineering professionals, pulled in $20 million a year.[5] Digital computing saw dozens of electrical engineering research assistants working on their master's degrees while the staff sponsored 21 doctoral theses.[6] Soon MIT was whistling a happier tune about its air defense research lab 12 miles out of town. Lincoln had become an essential part of MIT's educational mission. As one of its directors asked rhetorically: "What can Lincoln Laboratory do for engineering education? It can provide entirely new opportunities to advanced students who wish to work in the complex fields at today's technological frontier."[7] Today's technological frontier, no less. The bright boys and their technology were suddenly a well-accepted part of the establishment. There was no denying the fact that Whirlwind's DNA was everywhere at Lincoln; it dominated the environment; and everyone had their hands deep within its genetic code rearranging bases into new forms. There was, of course, the XD-1, then TX-0, TX-2, the radars and modems and communications and peripherals. It set off chain reactions of ideas that skittered down every corridor and rattled the door of nearly every laboratory.

J. C. R. Licklider was a case in point. Wes Clark, the builder of the TX-2, first met Licklider one evening in the basement of Lincoln. Licklider was trying to operate a console. Clark briefly introduced him to the TX-2, its display screen and programming techniques.[8] Licklider was hooked. He'd return later, either alone or with others, for more pointers from Clark. "What Licklider gained from these subterranean sessions was a profound conviction about the importance of interactive computing . . . which he later articulated in "Man-Computer Symbiosis" ."[9] A machine was influencing a man's thinking; the man, in turn, would later use that new-found thinking to revolutionize the machine, and human interaction with it. This was the enduring legacy of the bright boys' tool placed into the hands of the right person at the right time. The same TX-2 placed again into the hands of Ivan Sutherland brought forth a revolution in computer graphics with Sutherland's *Sketchpad* program (1963). Whirlwind's revolutionary idea of many consoles sharing the same computer, as in Room 222, would later galvanize young MIT researchers Fernando Corbato and Robert Fano into building a time-sharing computer system (1961). Time-sharing, they wrote, "can unite a group of investigators in a cooperative search for the solution to a common problem or it can serve as a community pool of knowledge and skill on which anyone can draw according to his needs."[10]

Decades before the personal computer, computing was getting personal and social. In an age when people's only interaction with a computer was to stuff stacks of IBM cards up a computer's rear and then wait, Whirlwind's DNA was doing magic. For Licklider, his time with TX-2's screen was a profoundly moving experience. Interactivity became his drumbeat, a driving force of the most personal and social kind, heralding, as one observer describes it: "the crown jewel of Licklider's crusade . . . his initiation of events leading to the Internet."[11]

In 1958, like a planetary probe heading into a new world, Clark sent the TX-0, slightly modified, winging off to MIT, because, as he put it, " . . . what MIT needed at the time was a computer that you could get your hands on. That was the one thing that they did not have."[12] On the receiving end were the likes of Jack Dennis, who used TX-0 to build his MACRO assembler, and Tom Stockham, who built the FLIT debugging program, both of which were firsts of their kind in computing. Another MIT "Tixo" alum, J. Martin Graetz, summed up his time with TX-0 as, "the chance to work on this computer was in many ways a rite of passage; it meant that I had joined the ranks of the real programmers."

New Voices

In 1955, new voices were heard in the corridors of Lincoln Laboratory—eight new voices fresh in from the West Coast, from Santa Monica and the RAND Corporation. Eight computer programmers arrived to partner with the bright boys. The XD-1 was scheduled to get a fresh new computer program and RAND was hired to do the job. "The Air Force selected RAND as the only logical choice for this activity [RAND was a Hap Arnold arrangement with Douglas Aircraft created to do research exclusively for the Air Force]. One reason: RAND had a corner on the country's programmers."[13] There were no more than 200 programmers in the country and ten percent were RAND employees. James Wong was one of them.

Wong graduated from UCLA in 1952 and went to work for RAND as an associate mathematician. RAND had an IBM 701 as well as its famous von Neumann clone, the Johnniac, for which Wong and Cecil Hastings developed programming for high-speed digital computers for RAND's System Development Division. RAND's job with XD-1 was to take over the operational program development, create supporting software, and then carry out the installation of that programming in all of the SAGE Direction Centers.

Wong's first time in the computer room was breathtaking and mind-boggling: "You're engulfed in there. You get a strange feeling that you are part of the computer, and the computer is part of you! It's like a member of the family."[14] He got acclimated by going around with the equipment team to check out long-range radars, gap-fillers, and the height finders. While hobnobbing around he met pockets of other contractors who, like himself, were cutting their teeth on the giant computer. Bell Labs, RCA, AT&T, Burroughs, and IBM had crews similarly "engulfed" in XD-1. The RAND contract called for 25 programmers, if RAND could find that many to hire. Eventually, between direct hires and training, RAND amassed 700 programmers, 300 of whom went to Lexington. The System Development Division got so bloated with people and revenue that RAND spun it off as the System Development Corporation in 1957.

Wong's first assignment was to write a test program for the ground-to-air data link, but his big test came when he got the assignment to lead his team in writing the machine's Program Executive Control (PEC), which today is a computer's operating system. The PEC turned out to be 230,000 operating instructions.[15] Wong's team did it all in assembly language, since there were no higher-level languages yet available. It took three weeks to complete. "In those days we had one card per instruction," and one evening at midnight they gathered and fed all their IBM cards into XD-1, hoping mightily that there were no errors and that the machine would eat every one with gusto. All worked out well. However, if any bugs cropped up, IBM offered a bounty to find and fix the errors; " . . . nominal awards were like $50 and $100, and it went up to $1,000 to $15,000."[16] The PEC was followed by utility and support programs that rung up an additional 870,000 instructions. In all, over 1 million lines of code went into what was then the largest program and largest programming job ever.

By the end of 1957, "the official handover of the software function from Lincoln to RAND was complete."[17] In 1958, Wong and 70 other programmers went to McGuire AFB, New Jersey, where for two years the blockhouse and two Q-7 computers awaited their coming. By June, the McGuire SAGE site was online, and with 40 programmers SDC did the same at Stewart AFB in New York. "Subsequent sites went on-line about every two months," with programming proficiency getting to the point where only 15 SDC personnel were necessary to crank each machine into operation.[18]

In 1998, at a lecture titled "Vigilance and Vacuum Tubes: The SAGE System 1956–1963," James Wong reminisced about those early days of his youth and the youthful profession of computer programming. For many, many years, he said, the word around the industry was: "It was first done in SAGE." "SAGE was the real-time, command and control computer-based system with capability so advanced," he told the audience, "that 40 years later, today, some of that capability can still be called state of the art . . . things like multiprocessing, real-time database management, distributed processing, time sharing, interactive displays, networking—they were all there in SAGE."[19] Wong and his team did all coding without the benefit of high-level programming languages. They were too early for the likes of FORTRAN, ALGOL, COBOL, and BASIC. Binary was the XD-1's native language; its arithmetic unit gobbled up with great alacrity machine code like this string: 00101111001011011000101101001110110111. But for Wong and his team, or any other programmer of the day, such strings of binary 0s and 1s were mind numbing. SAGE was programmed in machine language but had some shortcuts called "pseudo-instructions," what today are called "macros." Four such substitutes for binary were

- ETR for Perform an ADD to the Accumulator with a mask the size and position of the item;
- POS for Shift the accumulator so that the least significant bit of the item is in the least significant bit of the accumulator;
- RES for Shift the accumulator left so that the least significant bit of the accumulator moves to the least significant bit position of the item;
- DEP for Deposit into the word containing the item from the proper bit positions of the accumulator.

This concept of "pseudo-instructions" was implemented with the SAGE Communication Pool (Comm Pool), which was the central data definition used to assemble programs. To add two items and store in a third, the following sequence was used:

CLA ITEM1	Move ITEM1 to accumulator
ETR ITEM1	
POS ITEM1	
STO TEMP	Save ITEM1
CLA ITEM2	Move ITEM2 to accumulator
ETR ITEM2	
POS ITEM2	
ADD TEMP	Add the two items
RES ITEM3	
DEP ITEM3	

Such "pseudo-instructions" were less direct than the binary code that was SAGE's own language, but they were a lot more "programmer friendly" than 0010111100101101100010110100111011011011.

In 1958, Bell Lab's John Tukey gave it all a name: "software." During the next decade, the 1960s, software would take root and begin to flourish. A *Business Week* article published in 1964, *"New Tool—New World,"* crowned software as computing's "new tool," and predicted:

> As more and more human abilities are transferred into the computer through programming and thereby become part of a new kind of library of skills, the effect may be equal in kind to the change that occurred when written language appeared. Writing freed mankind from total dependence on memory and permitted the accumulation and selection of effective knowledge. The computer program in turn accumulates and preserves skills.
>
> There are enormous implications in this—for business, education, training, and future employment. There is not much point in having someone spend a great deal of time mastering the kind of skill that a computer can learn in a few seconds by having a program fed into it.[20]

Over time, software would inherit the earth by making relative simplicity from complexity, by creating human understandable computer languages that freed people from coding in machine language. "You needed an internal program that was smart enough and fast enough to translate a programmer's instruction into efficient machine code."[21]

Kicking off the hit parade of programming languages to come were John Backus and his handpicked, IBM team of programming misfits, who, after three years of working the problem, came up with the first permanent solution in April of 1957.[22] The question they pursued, was, as Backus put it, "Can a machine translate a sufficiently rich mathematical language into an efficiently economical program at a sufficiently low cost to make the whole affair feasible?"[23] Their answer was FORTRAN, an acronym for "FORmula TRANslating System." FORTRAN was not perfect first time out of the chute, but it was a great beginning. FORTRAN enabled programmers to write programs 500% faster, while the execution efficiency of translating FORTRAN code into machine language and executing commands suffered a reduction of only 20%.

Interestingly, shorthand English and algebraic formulas—the same method that FORTRAN used—were first used on Whirlwind in January of 1954. Backus, writing years later, recounts how J. Halcombe Laning, Jr and Neal Zierler had the first algebraic compiler running on Whirlwind. In the summer of 1954, Backus was at MIT for a demonstration that only confirmed for him the elegance and naturalness of using a concise mathematical language for FORTRAN.[24] The impetus for developing FORTRAN, as it was with many such programming languages of the time, was simple pragmatism. "FORTRAN did not really grow out of some brainstorm about the beauty of programming in mathematical notation; instead it began with the recognition of a basic problem of economics: programming and debugging costs already exceeded the cost of running a program, and as computers became faster and cheaper this imbalance would become more and more intolerable."[25] The door was open for programming to bubble and percolate throughout computing.

Exterior of first SAGE installation, McGuire Air Force Base, Long Island, New York, 1958. (Photo courtesy of MITRE Corporation.)

XD-1's brethren, the AN/FSQ-31, would get a language of its own called JOVIAL, but the language proved cumbersome and soon fell to the wayside. However, the breakthrough process that FORTRAN opened for others would be repeated with successes like COBOL, ALGOL, and BASIC. Sometimes programming languages took years to perfect and then were built out still further by others. In the 1970s, Dennis Ritchie of Bell Laboratories developed the programming language called C; ten years later, Bjarne Stroustrup, also at Bell Laboratories, sought to make writing good programs even easier and more pleasant for the individual programmer. Standing on the shoulders of C, he built a newer programming language called C++. C++, although based on C and retaining a great deal of its functionality, was actually a separate programming language.

Like their hardware pals, early programmers were an inventive lot. "The programmer had to be a resourceful inventor to adapt his problem to the idiosyncrasies of the computer: He had to fit his program and data into a tiny store, and overcome bizarre difficulties in getting information in and out of it, all the while using a limited and often peculiar set of instructions. He had to employ every trick he could think of to make a program run at a speed that would justify the large cost of running it. And he had to do all of this by his own ingenuity, for the only information he had was a problem and a machine manual."[26]

If Wong and his programming mates needed any respite from the rigors of taming XD-1, they could nip off a mere two miles away for an evening's diversion in the chummy confines of the Lexington Theatre. In 1957, the hot flick in town was the romantic comedy *Desk Set* starring Spencer Tracy and Katherine Hepburn. Strangely, right after the opening credits ran, there was some splashy billing for IBM, singling itself out for credit as the hardware provider and set designer for Spence and Kate's costar, a computer named EMERACK, the Electro-Magnetic Memory and Research Arithmetic Calculator, nicknamed Emmy. Tracey as Richard Sumner is Emmy's inventor and an MIT-trained efficiency expert. Hepburn playing Bunny Watson runs the corporate library. Emmy has arrived to "help" Watson's staff do their work faster and better; and Sumner is there to "help" ease the integration of his machine with Watson's very leery coworkers.

Actually, Watson's company bought two of Sumner's "big brains," one for research and the other installed in the payroll department. During the machine's first hellos with the staff, Wong and his buddies would have glimpsed a dawning in the future relationship between man and machine: the IBM opening credit was a sales pitch and also a public relations message to position computers as work assistants and not inhuman job jackers.

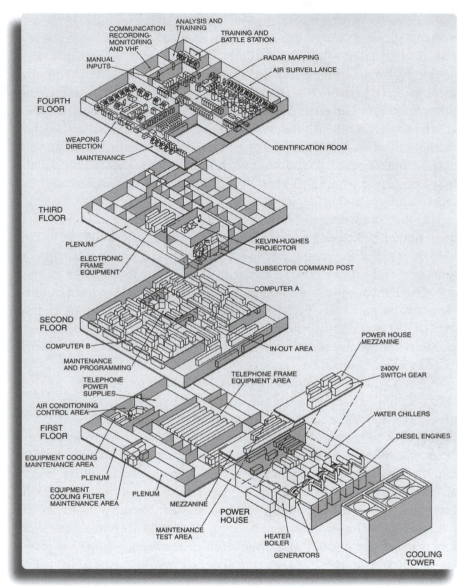

SAGE Direction Center schematic. (Photo courtesy of MITRE Corporation.)

And who better to costar with that kind of PR message than America's darlings, Spencer Tracy and Katherine Hepburn. IBM even allowed a humorous computer glitch to sneak into the script when the computer sends every employee a pink slip . . . even the boss. Computers were on the rise, automation was on the creep in America, and IBM was packaging a friendly smile on its product line. That same year in New York City there were 200,000 elevator operators opening and closing elevator doors and whisking passengers up and down buildings large and small. But not for long. Within a single decade nearly all would be out of work from the kind of job that would never ever come again. Machines would be smoothing the leveling of elevators at every floor and machines would be warning passengers: "Watch your step."

If Wong was at the Lexington Theatre in October, the newsreel before the feature film would have been more eye-popping than EMERACK. Inescapable was the shocking news of October 4th. The Soviet Union blasted Sputnik into orbit around the Earth, which had huge implications for an air defense system trained only to look for manned bombers. It would be some years before anyone figured out how to make a portly nuclear bomb light enough and svelte enough to stick atop a missile. And even more tricky, there was the larger problem of building a guidance system accurate enough to control a missile's flight thousands of miles to hit its intended target. As of 1961, the USSR had but six of these long-range R-7 missiles, whose radio-control flight accuracy was dubious at best.[27] The R-7 took too long to fuel, its above ground launch facilities were large and vulnerable to attack, and it could only be held on standby for 24 hours before the propellant seals began to fail.

Bombers were still very much the only real way of dropping killer blows on American soil. But that didn't stop the panic in the streets. And it also did not stop more than a few long faces in Building F from looking up at XD-1 and wondering, what next. President Eisenhower went on television and radio to reassure the American people that America's space program was second to none.[28] A Japanese newspaper called Sputnik "a Pearl Harbor for American science" while the British prime minister declared, "Never has the threat of Soviet communism been so great."[29] Two months after Sputnik, the Air Force got an Atlas rocket aloft; and on January 31, 1958, barely three months after the Soviet launch, the United States launched its first satellite, *Explorer I*. But the thunder had already been stolen by Sputnik, and American science and engineering turned into one huge, hyper catch-up program. And from kindergarten through to post-graduate studies, the American educational system was scrutinized, tinkered with, and pumped full of cash to outrace America's Cold War competition. Neil Armstrong's footprint on the Moon in 1969 was one of the direct results. Although most of the Sputnik-inspired fear syndrome was unfounded, everyone remained in panic mode for much of the following decade.

High-Tech Air Force

On the other side of the fence from Lincoln Laboratory was the 1,100-acre Hanscom AFB, which was hopping with activity from 1951 onward. By October of 1953, MEW Hill had been leveled to make way for new runways, two large hangars were built, a headquarters building—Building 1600— went up, as did a chapel, base housing, and the dedication of the Electronics Research Directorate, and the Research Services Division. Five months after XD-1 arrived, the AFCRL moved from Albany Street to Hanscom, joining up with aircraft from 6520th Test Support Wing. When XD-1 was ready to take over the network from Whirlwind, and then when XD-1 relinquished the network to the SAGE continent-wide Direction Centers, Hanscom AFB would be ready for both. The Air Force was going bigtime for high-tech electronics and doing it in a hurry.

Much of the behind-the-scenes work in creating this forward-looking Air Force originated with Muir Fairchild and Gordon Saville, especially Saville. Just before his retirement in the spring of 1951, when he was Deputy Chief of Staff for Defense, Saville hired Ivan Getting "to track new scientific and technological developments, and connect these with Air Force strategy and operations."[30] To assist him as deputy, Getting named a former scientific liaison officer, Colonel Bernard Schriever (who would eventually build and command the country's entire ballistic missile program). Saville had previously named General Donald Putt to be Director of R&D, and Louis Ridenour as Chief Scientist. All of these appointments would have far-reaching impact on the Air Force, which, barely ten years after its founding in 1947, had vaulted itself into prominence as the premier high-tech service in the US military.[31] When Getting resigned to take an executive position with Raytheon, Col. Schriever replaced him as Assistant for Development Planning; then in 1954, promoted to General, Schriever undertook the Air Force's most important R&D job, the Atlas ICBM missile program.

To build the missile system, Schriever aped the systems engineering approach and techniques learned from Getting and SAGE.[32] His systems integrator and Lincoln Laboratory-type facility would be Ramo-Woolridge, led by Simon Ramo and Dean Woolridge. The practice of "concurrency," whereby many manufacturers simultaneously made parts for a single project like the Atlas missile, was exactly the same as the relationship of SAGE's myriad manufacturers producing parts for the air defense system.[33] He would later codify it as the 375-Series of Systems Management regulations. Schriever even copied his famous weekly staff meetings called Black Sundays from the bright boys brand of engineering management, continuing on a line of succession from the Valley-era, Friday ADSEC meetings and from Forrester's weekly skull sessions first at the Digital Computer Laboratory and still later at Lincoln Laboratory.

Schriever, in selecting his own deputy, chose General Charles Terhune over General James McCormack, which set into motion two events critical to SAGE's future. McCormack, passed over for the Schriever position, retired from the Air Force to become an MIT vice president in charge of government relations. In 1958, it would be McCormack who eased SAGE out of a Lincoln Laboratory grown weary of SAGE's monopolizing presence, and who guided it to its next home, the MITRE Corporation. In June of 1960, Ivan Getting left his vice presidency at Raytheon, moving to Inglewood, California to head up the Aerospace Corporation, a wholly-owned Air Force subsidiary for advanced planning, technical evaluation, and systems engineering of Schriever's ballistic missile programs. Aerospace Corporation took over much of the former Ramo-Woolridge relationship with Schriever. General Terhune would soon turn up as the commander of the Electronic Systems Division (ESD) at Hanscom AFB, which by 1961 would fall completely under the sway of Schriever as part of the Air Force Systems Command.[34]

The bright boys' fingerprints for systems engineering, management, and outsourcing were everywhere in the new Air Force. Axed by the Navy in February of 1950 and scheduled for complete mothballing in June of 1950 as recommended by the "Report on Electronic Digital Computers by the Consultants to the Chairman of the Research and Development Board," the bright boys and Whirlwind had provided huge return dividends to their Air Force benefactor. In addition, both the Polaris missile program that put the Navy back into the national defense picture, and the NASA Apollo missions that put humans on the Moon, utilized Atlas-type systems engineering.[35] Each became the success that it was because of Atlas and SAGE. And if not for George Valley's telephone call to von Karman and Air Force money flooding the Barta Building, Schriever might not have systems engineered much of anything without hugely overspending in time and money. And with Sputnik hysteria in the offing, some fast, positive response would be absolutely necessary.

The Air Force did not gravitate to computing and high-tech electronics because it was clairvoyant, but rather out of necessity arising from being the Department of Defense's annointed defender of the nation's Cold War skies. To be such a defender required having jet bombers, supersonic interceptors, and a bristling array of ballistic, intercontinental, and surface-to-air missiles. And in turn, all of this demanded—for design, testing, manufacture, and operation—the accuracy, speed, and reliability of digital computers.

Hap Arnold and von Karman put the Air Force on the right course: what Arnold, as far back as 1938, had called "sowing the seeds" of avanced technology. Schriever was reaping the harvest of Arnold's sowing and at the same time sowing more seeds of his own. High technology was becoming an integral and inseparable part of Air Force culture. For the bright boys, the Air Force afforded them "maximum latitude and flexibility in the interpretation of contracts . . . fullest availability of military plans . . . and reasonable freedom to manage projects."[36] That was an unprecedented luxury in building a system that had so much national security riding on it. Bob Everett was surprised and grateful that the Air Force, whose reputation was riding on it all, was not terrified or more circumspect. "Those of us who were designing SAGE believed in it, and I don't know how we could have done the job if we didn't. I was amazed at the time and I'm still amazed at the unflagging support of the Air Force. Truly remarkable."[37]

Beginning in 1954, the year after the successful Cape Cod tests and the same year that the name SAGE began to take hold, technology began to transform the way the Air Force operated. Eisenhower was in office as of January of 1953, and he informed Congress of his New Look strategy, which relied on dominance in nuclear weapons, strategic air power, and the doctrine of "massive retaliation" to any Soviet threat against the United States and its NATO allies.[38]

Curtis LeMay, SAC commander since 1948, began making substantive changes to his forces to gear up for Eisenhower's New Look. Just as SAGE intended to disperse to 23 Direction Centers around the country, blockhouses for which were well underway, LeMay in 1956 similarly dispersed his SAC forces to 29 US and 10 overseas locations. Many of his US locations were curiously close to SAGE Direction Centers, sometimes elbow to elbow. Soviet bombardiers would most definitely relish the idea of hitting a SAGE blockhouse and a SAC base with a single bomb. But there was wisdom in LeMay's plan. SAGE Direction Centers were worth their weight in gold for their capabilities in gathering information, military information that might well be the difference between a bomber force getting airborne or being caught on the ground. The military had nothing remotely comparable to the SAGE network. Information meant time, and time would be the difference maker between retaliation and obliteration.

LeMay had sweated out the Nunivak scare of 1952. He remembered how both the Air Force and the event had ended up as sensational copy in *Flying*

Saucer Review. Time could not dull his recollection of how his SAC forces were caught on the ground and how the Air Defense Command had been criticized by the Pentagon for undue panic and overreaction. He knew that timely information was the key to readiness. He knew that Nunivak had laid bare the Air Force's information deficits of woefully poor radio communication and slow, leased Teletype circuits. He dumped all the slow stuff in 1956; dug a huge, underground command post of reinforced concrete at SAC headquarters at Offutt AFB, Nebraska; and ordered in a primary network for the transmission of Emergency Action Messages (EAMs) to his SAC bases, which later expanded into the 465L system that included one of Whirlwind's descendants, the AN/FSQ-31, plus a viewing system called a "Quadrajector." Because it was near impossible to train airmen to react to massive Soviet bomber attacks—when there actually weren't any—RAND created for the Air Force the first computer program that simulated such bomber strikes. With SAGE's live radar shut off and hundreds of fake inputs fed into the Q-7 computer, the Air Force simulated reacting to a Soviet attack and practiced at getting SAC forces airborne in a hurry.[39]

LeMay early on realized the necessity of SAGE information. He had been at Maxwell Field in the late 1930s to witness the infighting between Chennault and Saville with the bomber mafia. LeMay was a mafiosi. He had sided with the nonsense of unescorted bombers and high-altitude daylight bombing that the Kammhauber Line of flak towers and the *Luftwaffe* had flamed into near extinction. It was all too distinct for LeMay to recall the B-17 raids over Germany, Bloody Thursday, and the void of information at 30,000 feet. He and Vandenberg were saber-rattling buddies from way back, but they were not fools. SAGE might be an air defense umbrella over the continent, but for SAC it was a weapon. It was a formidable information weapon that harkened its clarion call back through the ages, right to Sun Tzu's *The Art of War*—information wins wars!

LeMay was hell-bent to get his fair share and more of it, and he understood that, "The survival of strategic aircraft on a given air base was related to the degree of alert practicable and the warning time available. With the Distant Early Warning (DEW) Line in operation against Soviet jet aircraft, LeMay counted on getting two hours' tactical warning time and believed that it would be possible to get something like 60 percent of his aircraft into the air in this time. Against Soviet ICBM attack, however, the zone of interior bases could count on only about a 15-minute tactical warning . . . "[40] Well before Soviet bomb bays opened, LeMay intended for his B-47s and B-52s to be long gone. DEW line radar data would be flowing into SAGE Direction Centers and networked throughout the system . . .

including right into Offutt's big computer. Information about weather conditions; information from radars south of the DEW line, picket ships, Texas Towers, and early-warning aircraft would be fed to the blue-screened situation room where LeMay and his command staff could observe all the radar tracks from any incoming bogey. All of it in real time! SAGE, integrated with Nike and BOMARC missile sites, also provided real-time status reports on US defensive weapons.

When Soviet bomber technology produced the faster, higher-flying Tupolev-16 Badger and the Tupolev-95 Bear, US F-86s and F-89s were left flatfooted to offer chase or climb to ceilings like 46,000 feet and above. Saville's interceptor program produced the delta-wing F-102 "Deuce": the first supersonic military jet at Mach 1.2 with a ceiling of 55,000 feet and a thousand-mile range. Pilots depended on the information flow to their cockpit displays from the SAGE data link. Even better than the Deuce was the F-106 Delta Dart, called the "Six" by its pilots. The F-106 had an on-board digital computer that when hooked up with SAGE allowed SAGE to select and fire weapons as well as autopilot the craft.

Most pilots preferred to do their own trigger pulling. "All you had to do was select the armament, shoot and return to base," recalls William Neville, a former F-106 flyer. "The computer locked onto the inbound target and onto you. You saw everything on the tactical situation display."[41] The Canadian Air Force opted for the twin-engine, Mach 1.7 F101 Voodoo, also SAGE compatible. Both the F-106 and the Voodoo with AIM-4 missiles and nuclear-tipped Genie missiles, connected to data flows from SAGE, shut the door to enemy bombers coming over the Arctic Circle. LeMay remarked that his new *gladius informatiensis* had made "the fundamental concept of a coordinated air battle and a defense in depth a practical reality."[42]

On June 28, 1958, the enlightened commander of 1,500 jet bombers came to McGuire AFB to honor his new information weapon. It was the opening of the New York Air Defense Sector, and the dedication of the first SAGE Direction Center. "SAGE does not think," LeMay told the crowd at the ceremony. "It gathers and stores information and presents a picture on which man can act. SAGE does not nullify the need for well-trained and proficient personnel. It enables such personnel to do a better job."[43] LeMay's understanding of the new tool was converging with Licklider's.

Information Gets Technology
From the Barta Building to Lincoln Laboratory to a national grid of 23 Direction Centers (and another at North Bay, Canada), the bright boys' technology was spreading. With a dozen commercial computer and electronics

firms contracting on the SAGE project, it was only a matter of time until the technology was refined, repackaged, and sold as commercial products. It was only a matter of time as well for thousands of newly trained digerati to leave these same contractors for opportunities elsewhere. System Development Corporation was one of many that witnessed the outflow. In *The System Builders: The Story of SDC*, Claude Baum follows the SDC diaspora of SAGE-trained programmers as they migrated from SDC to start companies or meld into corporate programming staffs. Many drifted into then-quaint, backwater California towns like Palo Alto and Mountain View, getting ready, in many cases, for guys like Robert Noyce to hit town with cool ideas about microprocessors.

Moving information into the arms of technology could not be accomplished without government intervention. Individuals, universities, and corporate research laboratories simply weren't enough. Bright boys' technology took lots of government cash and government faith. That trend would continue. An important difference post-1957 was that SAGE's success gave the government confidence in technology; it believed that technology could come to the rescue with solutions for great national problems. Had XD-1, the SAGE network, and continental air defense become colossal failures, government reluctance to throw money at a potential technological breakthrough would have been more than obvious.

Faith in American technology to meet the threats posed by the 1957 tests of Sputnik and the Soviet Sapwood ICBM bolstered the government's resolve to let the money flow. With a worldwide recession in full bloom from April of 1957 to August of 1958, and US unemployment rates skyrocketing, government spending was under a microscope. Although faced with a financial crunch and the howlings of Democrats looking to sweep Republicans out in the 1960 elections, Eisenhower spent. The old general in him was eyeing worldwide events and not the West Wing. NASA, the National Aeronautics and Space Administration, was formed from the old ribs of NACA, the National Advisory Committee for Aeronautics; the US Interstate Highway System was begun; and a little-known, specialty agency was formed called ARPA (the Advanced Research Projects Agency). ARPA would become the engine that pushed the bright boys' technology the final mile to modern Information Technology and the Internet.

Intercontinental ballistic missiles and anti-ballistic missiles were what was on Ike's mind at the forming of ARPA.[44] Well before Sputnik and Sapwood, US intelligence had been aware of and watching Soviet missile firings. A huge, top-secret radar at Samsun on the Black Sea coast of Turkey monitored Soviet intermediate-range and long-range test firings into the central Asian desert, and U-2 spy plane flights in June of 1956 photo-graphed the facilities in Kazakhstan.[45] The White House recognized that a single, concerted Soviet effort in missile technology had accomplished much, whereas in the United States, the rivalry among its three services had diluted missile development. In his 1958 State of the Union speech, Eisenhower promised to bring a halt to the rivalry.

Just before the Special Committee on Space Technology was to meet in February of 1958 to decide, among other things, the fate of NACA, basic missile research, and who among the services would be the heavyweight of the missile program, Eisenhower and his Secretary of Defense Neil McElroy initiated Department of Defense directive 5105.15, establishing the Advanced Research Projects Agency. The directive gave ARPA a deep reach into "the direction or performance of such advanced projects in the field of research and development as the Secretary of Defense shall, from time to time, designate by individual project or by category." With the Air Force and its Atlas, Thor, and the Titan missiles; the Army with Jupiter; and the Navy with Polaris, Eisenhower looked to ARPA to reign in some semblance of control.[46] If the idea of an anti-ballistic missile seemed fea-sible, what was then joked about as "trying to hit a bullet with a bullet," then Ike liked ARPA for that role as well. Another of the charges from di-rective 5105.15 was "to think independently of the rest of the military and to respond quickly and innovatively to national defense challenges." Bold projects that advanced America's defense-related technologies got prece-dence. And that is where computers slipped into the mix.[47] A year on and NASA would scoop up all the missile business, leaving behind for ARPA the visionary, blue-sky technologies associated with computing.[48]

Wes Clark's TX-2 was certainly blue-sky enough. His TX-2 became fa-mous for "advanced graphic display research;" and the TX-0 that he and Bill Papian sent to MIT became "famous as the favorite tool of *hackers* in Building 26." Since all computers of the time displayed information via Teletype only, graphic displays were of particular interest to ARPA. The idea was how "SAGE-like displays might be adapted to many types of com-puters, not just the big ones used to monitor air defenses."[49] Sutherland's *Sketchpad* was just such a graphics system that could be of great interest to ARPA. In fact, *Sketchpad* was more than just a display. It could store vi-sual patterns as easily as alphanumeric data; and as such was a "simulation language that enabled computers to translate abstractions into perceptu-ally concrete forms."[50] The Building 26 hackers would qualify their blue-sky credentials when they used their TX-0 to become the core group who pioneered time-sharing.

Display graphics and time-sharing were two other large helpings of bright boys' technology poised to get exposure in the greater research community, joining keyboard input, magnetic memory, and the modem as essential bright boys' originals going public. Of course, the real must-have companion for graphical input/output and time-sharing was Whirlwind's specialty: real-time computing—the linchpin for making the flow of information really worthwhile. No one was going to wait for information to meander to a screen. The essence of a screen was to see it now, which meant to see it fast.

In 1957, two months before Sputnik, the National Office Management Association conducted a survey of almost 4,000 US firms with 5,000 office workers or more. The survey found that 50% had already installed a mainframe computer (each with a purchase price of over a million dollars), while another 14% were waiting for the delivery of theirs. The vast majority of the work asked of these computers was payroll.[51] These 2,500 companies, with an aggregate of 12.5 million workers, were spending $2.5 billion to automate only their payrolls. Hollywood was correct: business looked exactly like *Desk Set*. The aspect of time-shared, real-time computing and graphical displays entering *Desk Set* environments all over America would be nothing short of revolutionary. These new tools, if they could edge their way in, would change the way information was processed, and as Licklider had predicted, they would "also change the way people thought."[52]

Importantly, businesses had taken serious notice of computing and were investing in the technology. However, the going was slow. Grappling with card-fed computers was awkward. "Computers worked like machines on an assembly line—repeating the same operations as records were passed through them one at a time."[53] An application like payroll might well be split into ten or twenty runs, e.g., deducting union dues from an already calculated payroll required a separate run. It was far faster and more efficient than payrolls worked by hand in rooms full of office workers with calculators, but it was still slow and cumbersome. This was a long way from the man/machine interaction already at hand with SAGE.

Needed as well as new computing tools was the realization that there was a "relatedness" to information the parts of which could be brought together to form a picture of overall business operations. As in LeMay's computer battlescape, where a computer "gathers and stores information and presents a picture on which man can act," so too could computers gather together information to present a picture upon which a business could act.

Business had no Barta Building in which to incubate and test; it would have to wait, borrow, and adapt both equipment and theory as the computers worked their way slowly into each business operation.

The dawning came slowly and in waves. "In 1953, during what was probably the first extended discussion of information as an abstract quantity to reach a large executive audience, *Fortune* magazine lauded it as a great and almost unknown scientific theory whose impact on society was likely to exceed that of nuclear physics."[54] The article, "The Information Theory" by Francis Bello, focused on the technical and electronic communications aspects of Shannon's new theory of 1948.

New terms like "Automation," "Electronic Data Processing," and "Management Information Systems" began cropping up in management texts and journals, each with its own coterie of theorists and consultants pushing the new ideas into the workplace. Howard Levin's 1956 *Office Work and Automation* called for a new breed of information specialists— information engineers—as well as a vice president for information to improve and refine the effectiveness of analyzing and using information. Putting together computers, information, automation, and management, two University of Chicago Business School professors in 1958 came up with the term "information technology." Their article in *Harvard Business Review* titled "Management in the 1980s," predicted a future workplace with corporations transformed by computer hardware, operations research methods, and simulation programs.[55]

By far the most inspirational and clearly articulated look at computers was delivered by Edmund Berkeley in his 1949 national bestseller *Giant Brains, or Machines That Think*. Berkeley was the first to introduce electronic computers and their potential use in business to a general audience. "Berkeley . . . gave early expression to the idea of information as a ubiquitous presence in the natural and social worlds. He made the computer less threatening by presenting it as the latest and most powerful in a series of pieces of 'physical equipment for handling information' that included everything from nerve cells to human gestures."[56] "In language that managed the delicate trick of being exquisitely clear and uncompromisingly evangelistic, Berkeley described how a computer works, step by step, instruction by instruction. Employing numerous diagrams, and painstakingly explaining every underlying concept (like "binary" or "register" or "input/output") as if it had never been explained before, Berkeley demonstrated how it was possible to move digital information from one "place" to another—and how a set of on/off switches, if wired correctly, could perform operations on that information, handling such extraordinary feats as the addition of two plus two."[57] In 1947 while working for the Prudential Life Insurance Company, Berkeley was the guy who engineered his company's first computer purchase: a UNIVAC, which Eckert and Mauchly were unable to deliver before their company's collapse.

Drinking in every delicious page of Berkeley was a 25-year-old named Doug Engelbart. Five years out of the Navy, directionless, drifting in and out of jobs on the periphery of computing, Engelbart was the exact right guy to profit from the Moore School lectures, but never went. If the bright boys had a spiritual and intellectual lost member of the gang, that unique spirit was Doug Engelbart, future creator of both the modern concept of the graphical user interface (GUI) and the GUI's illustrious sidekick, the mouse. Engelbart was more than just an inventor of "things" that made computing easier and more intuitive; he wanted to use computers to help people to learn. In *Tools for Thought*, Howard Rheingold devotes a chapter to this curious man's journey to recognition, titled "The Loneliness of a Long-Distance Thinker." "Loneliness" and "long-distance thinker" are key words for a young man with ideas aplenty stuck on the West Coast where he was looked upon as a crackpot by hostile colleagues.[58]

At 211 Massachusetts Avenue, Engelbart would have fit right in. As Douglas Ross had before him, Engelbart would have gravitated to Room 222 for the epiphany of a lifetime. He would have looked at Bob Everett's light gun and seen the makings of his mouse; he would have seen in Bob Wieser's time-shared consoles his learning system for computer interaction with people; he would have seen in Forrester's dogged pursuit of magnetic RAM his own stubborn will to succeed against all odds; and, with the camaraderie of the bright boys for support, he too would have smiled at the vultures circling above the Barta Building. "I confess that I am a dreamer," said Engelbart years later. "Someone once called me *just a dreamer*. That offended me, the *just* part; being a real dreamer is hard work. It really gets hard when you start believing your dreams." In 1951 in San Francisco, no one believed in this Oregon farm boy and World War II radar operator's vision of the close interaction of computers with people.

In Sputnik-laden 1957, Engelbart got a job in Menlo Park, California at the Stanford Research Institute where research was ongoing into the scientific, military, and commercial applications of computers. It seemed a perfect match. But strangely and sadly for Engelbart, his ideas on computers interacting with people to "augment their intellect" did not fit in at all. "How many people have you already told?" questioned a colleague after hearing Engelbart's theories. "You're the first," he said in reply. "Good, now don't tell anyone else," came the surprising rejoinder. "It sounds too crazy. It will prejudice people against you."[59]

Curtis LeMay would have had full belief in Engelbart. But how was Engelbart to know that? Fortunately, the Air Force found him first. The Air Force Office of Scientific Research gave him the go-ahead and a small grant. The grant offered him the opportunity to freely explore the field. "It was lonely work, not having anybody to bounce these ideas off, but finally I got it written down." What Engelbart "got written down" for Air Force contract AF49 (638)-1024 was a remarkable, 139-page paper titled, *Augmenting Human Intellect: A Conceptual Framework*. This manifesto for "augmenting man's intellect" with "high-powered electronic aids" would lead to the development of computer-based technologies for manipulating information; in short, Engelbart was putting the "personal" into "personal computing." Upon its publication, ARPA took close notice, and soon Engelbart found himself pulled into the protective ARPA force field set up by Licklider and Wes Clark. Engelbart read Berkeley's book in 1950 but did not produce *Augmenting Human Intellect: A Conceptual Framework* until 1962. That's 12 years of laboring in the vineyards before his salvation arrived. That's a long time to wait. However, if the event chain leading from Whirlwind to the Barta Building, to George Valley, to the Air Force, and then to ARPA had not happened, Engelbart and many others may well have labored obscure and anonymous for a very long time. Engelbart would indeed labor on for years more, but not in anonymity.

The first glimmer of Information Technology and its rise to prominence in the business world came in 1961 as IBM's SABRE system (SAGE's business-world alter ego) pushed to completion. The brainchild of Perry Crawford, SABRE was first pitched to American Airlines in September of 1957, at which time the airline was quick to see both the critical importance to reservation bookings as well as the financial windfall that would usher up from such control. SABRE was a real-time, time-shared system where travel agents at consoles individually probed the central computer in order to view airline flight schedules and to book reservations. It was like a civilian-style Room 222. James Gallagher's 1961 *Management Information Systems and the Computer* makes a case study of the SABRE system that "has been a textbook example of the strategic use of computers ever since."[60]

And that was just the beginning. "During the mid-60s, computer makers cemented their commitment to the new vision of real-time, on-line, managerially oriented systems."[61] Behind the façade of this new vision, people and technology would move from mainframe computer to mini-computer to personal computing. The bright boys had poked a hole into the dome of Information Technology and the stars of IT came tumbling down. For the bright boys, their job was over; it was for others to crack ever-larger holes in the dome. Engelbart would be but one of many.

ARPA in 1962 would go one step further. Under its then director Jack Ruina, ARPA hired Licklider to run its newly formed Information Processing Techniques Office (IPTO) to "extend the research carried out into computerization of air defense by the SAGE program to other military command and control systems."[62] In particular, IPTO would refine SAGE's wide-area computer network to build a survivable electronic network to interconnect three vital Department of Defense sites: the Pentagon, Cheyenne Mountain in Colorado, and LeMay's SAC headquarters. One of the IPTO research grants produced the memorable and important Larry Roberts TX-2 to Tom Merrill FSQ-32 digital packet exchange over dial-up telephone lines in 1965, which triggered the beginning of ARPANET, which later evolved into the Internet.

SAGE had and still has its share of detractors. Some critics labeled SAGE as extravagantly expensive and a failure, rendered obsolete by Soviet ICBMs before a single SAGE Direction Center ever went into action. Closer scrutiny proves otherwise. ICBMs did not become the threat of choice overnight; it would take time before bombers would play second fiddle to missiles. And very quickly did SAGE descendants become the eyes used to search for those ICBM launchings continents away.

Every modern military, friend or foe, uses a command and control system directly descending from the SAGE lineage. Every modern air traffic control system guides and lands commercial aircraft using methods pioneered by SAGE. Today's computing business owes a huge debt to SAGE innovations, and should be especially grateful for the billions of dollars of revenue scuttling throughout the electronics industry because of SAGE ingenuity. SAGE was actually quite a bargain; it was peerless in buying peace of mind for millions of people and their governments. For a few pennies a day per person, every man, woman, and child in the United States and Canada, not to mention much of Western Europe as well, depended on this bargain-basement insurance policy. It kept adversaries unsure and off balance, which was certainly far cheaper than a war.

Although none exists, one would half expect some tribute to be in evidence somewhere for the bright boys. Maybe a large bronze somewhere along the Charles where Massachusetts Avenue meets the river, maybe something with all the boys cavorting together. Something the sun could glint over and play upon showing off spirited young men forever young.

Cocky looks with wry smiles of confidence from youthful faces who knew that they were working on the biggest of big jobs and that everyone depended on them. Maybe a life-size chunk of glittering statuary memorializing their great challenge, their intellectual curiosity, their courage, their bravado. Maybe something that is inspirational as well, something that a future Doug Engelbart in passing by might take heart from. But, sadly, no dice.

The bright boys have disappeared, seemingly erased. And their old haunt, the Barta Building, is still not a national historic landmark with commemorative plaque and slick visitors' brochure. Whirlwind was carted away in 1959 by programmer Bill Wolf of Lincoln Lab for use in his own consulting company, Wolf Research and Development. Decades later, pieces of the great machine turned up at computer museums and at the Smithsonian. Forrester and Everett turned up in 1989 at the White House for some long overdue kudos from George Bush, the elder. Each received the National Medal of Technology some 30 years after the fact.

The only remaining witnesses at 211 Massachusetts Avenue are the gargoyles that encircle the building. They've seen it all. And just maybe the best tribute to the bright boys are the rumblings 30 feet beneath Massachusetts Avenue as the Red Line subway clatters in and out of Kendall Square station every seven minutes. It was the bright boys who showed the world that computers could do more than just count. Every seven minutes in Kendall Square station the proof rolls in and out again.

Notes

Chapter One

1. Thomas Haigh. "Inventing Information Systems: The Systems Men and the Computer, 1950–1968." *Business History Review* 75 (Spring 2001): p. 12.

2. Charles Seif. *Decoding the Universe: How the New Science of Information is Explaining Everything in the Cosmos, from Our Brains to Black Holes.* New York: Viking/Penguin, 2006, p.3.

3. Paul Edwards. "Vigilance and Vacuum Tubes: The SAGE System 1956–1963." Speech transcript: Computer Museum History Center, Mountain View, CA, May 19, 1998, p.10.

4. Bruce Berkowitz. *The New Face of War: How War will be Fought in the 21st Century.* New York: Simon and Schuster, 2003, p.14.

5. Kenneth Schaffel. *The Emerging Shield: The Air Force and the Evolution of Continental Air Defense 1945–1960.* Washington, D.C.: Office of Air Force History, United States Air Force, 1991, p. 67, citing Preliminary Rand Report, subj.: "Active Defense of the United States against Air Attack," July 10, 1947, revised and reissued February 5, 1948.

6. *Review of the U.S. Department of Defense Air, Space, and Supporting Information Systems Science and Technology Program.* Washington: National Academy of Sciences, 2001, p.27.

7. Donald MacKenzie. *Inventing Accuracy.* Cambridge, MA: MIT Press, 1993, p. 28.

8. Robert Everett. "Whirlwind," in Nicholas Metropolis, J. Howlett, and Gian-Carlo Rota, eds., *A History of Computing in the Twentieth Century.* New York: Academic Press, 1980, p. 384.

9. Thomas Hughes. *Rescuing Prometheus.* New York: Pantheon, 1998, p.176.

10. Haigh, "Inventing Information Systems," p. 6.

11. Dik Daso, (Maj. USAF). "Origins of Air Power: Hap Arnold's Command Years and Aviation Technology, 1936–1945." *Air and Space Power Journal*, Vol. XI, No. 3 (Fall 1997): Maxwell Air Force Base, Ala.: Air University Press. 1996, p. 97.

12. Jay Forrester and R. R. Everett et al. "Forecast for Military Systems Using Electronic Digital Computers." Memorandum L-3, Servomechanisms Laboratory. Cambridge, MA: Massachusetts Institute of Technology, 1948.

Chapter Two

1. Sidney Shalett. "First Atomic Bomb Dropped on Japan." *New York Times*, August 6, 1945, p.1.

2. Joel Carpenter. Nuclear Connection Project. "UFOs & Nuclear Weapons." June 2002, http://www.nicap.org/ncp/nn-020628.htm. (accessed November 02, 2009).

3. "U.S. Economy 1952." *Time*, November 17, 1952, p. 33.

4. Stephen P. Moeller (Col.). "Vigilant and Invincible," *Air Defense Artillery Magazine*, May–June 1995, Chapter Two, p.7.

5. *Time*, "U.S. Economy 1952."

6. Ibid.

7. Carpenter, "UFOs & Nuclear Weapons."

8. Ivan A. Getting. Oral history conducted by Frederik Nebeker, IEEE History Center, Rutgers University, New Brunswick, NJ, 1995, p. 4.

9. George Valley. "How the SAGE Development Began." *Annals of the History of Computing*, Vol. 7, No. 3, July 1985, p. 204.

10. Guy Stever. *In War and Peace: My Life in Science and Technology*. Washington, D.C.: Joseph Henry Press, 2002, pp. 92–93.

11. Carpenter, "UFOs & Nuclear Weapons."

12. Paul N. Edwards. *The Closed World: Computers and the Politics of Discourse in Cold War America*. Cambridge, MA: MIT Press, 1996, p. 22.

13. Ibid.

14. Wayland Mayo. "Russian Aviation: Soviet B-29, A Tupolev Tu-4 Story." http://www.b 29s-over-korea.com/shortstories/russianclone.htm.

15. Valley, "Sage Development," p. 199.

16. Ibid., p. 198.

17. Ibid., p. 214.

18. Stever, *In War and Peace*, p. 90.

19. Ibid., p. 91

20. Carpenter, "UFOs & Nuclear Weapons."

21. Stever, *In War and Peace*, p. 89.

22. Ibid., p. 90.

23. John J. Magee. "High Flight," 1941. http://hopeeternal.wordpress.com /2007/11/10/high-flight-john-gillespie-magee-jr-1922-1941/

24. Thomas Hughes. *American Genesis: A History of the American Genius for Invention*. New York: Penguin Group, 1990, pp. 124–125.

25. Edwards, *The Closed World*, p. 11.

26. Joseph Baugher. "Russian and Soviet Nuclear Forces: Tu-4 BULL." http://home.att.net/~jbaugher/uscombataircraft.html

27. Ibid.

28. Getting, oral history, p. 4.

29. Moeller, "Vigilant and Invincible," p. 16.

30. Lynden T. (Bucky) Harris. *The DEWLine Chronicles*. http://www.lswilson.ca/dewhist-a.htm

31. Ibid.

32. Paul Bracken. "Warning and Intelligence." Chapter 2. *The Command and Control of Nuclear Forces*. New Haven, CT: Yale University Press, 1983, p. 16.

33. Stan Augarten. *Bit by Bit - An Illustrated History of Computers*. New York: Ticknor and Fields, 1984, p. 196.

34. Maurer Maurer. *Aviation in the U.S. Army: 1919–1939*. Washington: Office of Air Force History, 1987, p. 355.

35. Ibid.

36. Major Hugh C. Severs, (USAF). "The Controversy Behind the Air Corps Tactical School's Strategic Bombardment Theory: An Analysis of the Bombardment vs. Pursuit Data between 1930–1939." Research Report No. 97-0126. Maxwell Air Force Base, Alabama: Air Command and Staff College, 1997, p. 9.

37. Sever, *In War and Peace*, p. 4.

38. James D. Perry. "Air Corps Experimentation in the Interwar Years: A Case Study." *Joint Force Quarterly*, No. 22, Summer 1999, p. 47.

39. Ibid, p. 48.

40. Richard P. Hallion. "Control of the Air: The Enduring Requirement." Air Force History and Museums Program. Bolling Air Force Base, Washington, D.C., September 8, 1999, p. 15.

41. "General Carl 'Tooey' Spaatz." Foxhole Profiles, 2004, p. 11. http://www.freerepublic.com/focus/f-vetscor/1074257/posts

42. Thomas S. Kuhn. *The Structure of Scientific Revolutions*. Chicago: University Press, 1996, p. 5.

43. Ibid., pp. 16–17.

44. Fritjof Capra. *The Web of Life: A New Synthesis of Mind and Matter*. London: Flamingo, 1997, p. 6.

45. Kuhn, *Structure of Scientific Revolutions*, p. 6.

46. Gen. Gordon Saville (Maj. USAF-ret.). Interview by Thomas A. Sturm. United States Air Force Oral History Program, March 26–29, 1973, p. 62.

47. Sever, *In War and Peace*, p.1.

48. "The Bombing of Guernica." PBS Online. http://www.pbs.org/treasuresoftheworld/guernica/glevel_1/1_bombing.html

49. "The SCR-268 Radar." *Electronics*, September 1945, p. 100.

50. Kenneth Schaffel. *The Emerging Shield: The Air Force and the Evolution of Continental Air Defense 1945–1960*. Washington, DC: Office of Air Force History, 1991, p. 16.

51. W. F. Craven and J. L. Cate. *The Army Air Forces in World War II*. Volume I: "Plans and Early Operations: January 1939 to August 1942," Chapter 4: "The Air Corps Prepares for War, 1939–41." Chicago: The University Press of Chicago, 1950, p. 112.

52. Ibid, p. 107.

53. "The First Coup Attempt: September 28, 1938." http://www.nizkor.org/ftp.cgi/imt/nca/nca-01/ftp.py?imt/nca/nca-01//nca-01-09-aggression-04 (accessed November 02, 2009).

54. Greg Goebel. "Radar and the Battle of Britain." 2002. <http://www.vectorsite.net/ttwiz_01.html#m2>

55. Ibid.

56. Ibid.

57. Ibid.

58. J. E. Beasley. "OR-Notes" (course notes for Operations Research and Management Science) Department of Mathematical Sciences, Brunel University, West London. 2004, pp. 3–4.

59. Ibid.

60. Goebel, "Radar and the Battle of Britain."

61. Ibid.

62. Stever, *In War and Peace,* p. 25.

63. Ibid.

64. Ibid., p.39

65. Ibid., p. 34.

66. Ibid..

67. Saville, interview, p. 66.

68. Maurice V. Wilkes. "The Origins and Growth of Electronic Engineering—A Personal View." London: Computer Laboratory, University of Cambridge, 1993, p. 5.

69. Stever, *In War and Peace,* p. 26.

70. John Hendry. *Innovating for Failure: Government Policy and the Early British Computer Industry.* Cambridge, MA: MIT Press, 1989, p. 168.

71. Donald MacKenzie. *Inventing Accuracy.* Cambridge, MA: MIT Press, 1993, pp. 31ff.

72. Stever, *In War and Peace,* p. 29.

73. David F. Winkler. *Searching the Skies: The Legacy of the United States Cold War Defense Radar Program.* Langley, VA: USAF Headquarters Air Combat Command, 1997, p. 4.

Chapter Three

1. Thomas Haigh. "Inventing Information Systems: The Systems Men and the Computer, 1950–1968." *Business History Review* 75 (Spring 2001): p. 12.

2. Michael S. Mahoney. "The History of Computing in the History of Technology." *Annals of the History of Computing*, October, 1988, p. 113.

3. Haigh, "Inventing Information Systems," p. 12.

4. Kenneth Flamm. *Creating the Computer: Government, Industry, and High Technology.* Washington, D.C.: The Brookings Institute, 1988, p. 55.

5. Mina Rees. "The Computing Program of the Office of Naval Research, 1946–1953." *Communications of the Association of Computing Machinery*, Vol. 30, No. 10, October, 1987, p. 842.

6. Stan Augarten. *Bit by Bit - An Illustrated History of Computers.* New York: Ticknor and Fields, 1984, p. 2.

7. Jay Forrester. "The Beginnings of System Dynamics." Banquet Talk: Systems Dynamics Society: Stuttgart, Germany. July 13, 1989, p. 3.

8. Jay Forrester. "Lincoln Laboratory, MIT Historical Comments." Heritage Lecture Series. *50th Anniversary of Lincoln Laboratory.* November 26, 2001, p. 1.

9. Ibid., Forrester, p. 1.

10. Dale Kreiger. "The Story that Doesn't Compute." *Johns Hopkins Magazine*, November, 1999, pp. 5–6.

11. *New York Times*, February 16, 1946, p. 17.

12. Noam Nisan and Shimon Schocken. *The Elements of Computing Systems.* Cambridge, MA: MIT Press, 2005, p. 79.

13. Jay Forrester. "Whirlwind's Success." Transcript of talk from symposium: *Whirlwind's Genesis and Descendants.* Boston: Computer Museum, October 18, 1987, p. 2.

14. Augarten, *Bit by Bit,* pp. 197–198.

15. Stuart Bennett. "A Brief History of Automatic Control." *IEEE Control Systems.* June, 1996, p. 20.

16. Steve Bragg. "The SCR-584: Radar Tribute Page." www.hamhud.net/darts/scr584.html.

17. Ivan A. Getting. Oral history conducted by Frederik Nebeker, IEEE History Center, Rutgers University, New Brunswick, NJ, 1995, pp. 2–3.

18. Ibid., p. 2.

19. Robert Everett. "Whirlwind." in Nicholas Metropolis, J. Howlett, and Gian-Carlo Rota, eds., *A History of Computing in the Twentieth Century.* New York: Academic Press, 1980, p. 366.

20. Getting, oral history, p. 3.

21. Guy Stever. *In War and Peace: My Life In Science and Technology.* Washington, D.C.: Joseph Henry Press, 2002, pp. 38–39.

22. Ibid., p. 306.

23. Jay Forrester. "Lincoln Laboratory, MIT Historical Comments." *Heritage Lecture Series.* November 26, 2001, p. 5.

24. Ibid., p. 4.

25. Ibid., p. 6.

26. Stever, *In War and Peace,* p. 62.

27. George Valley. "How the SAGE Development Began." *Annals of the History of Computing,* Vol. 7, No. 3, July 1985, p. 218.

28. J. C. R. Licklider. "Man-Computer Symbiosis." *IRE Transactions on the Human Factors in Electronics,* HFE-1, March, 1960, p. 6.

29. Howard Rheingold. *Tools for Thought: The History and Future of Mind-Expanding Technology.* Cambridge, MA: MIT Press, 2000, p. 108.

30. Herman Goldstine. "Computers at the University of Pennsylvania's Moore School, 1943–1946." *Proceedings of the American Philosophical Society,* Vol. 136, No. 1, 1992, p. 75.

31. Richard Halstead-Nussloch. "Computers and Society: Professional Practices and Ethics, Lecture 6—Five Times One Thousand." Marietta, GA: Southern Polytechnic State University, 2001.

32. William Aspray. "Was Early Entry a Competitive Advantage? U.S. Universities that Entered Computing in the 1940s." *IEEE Annals of the History of Computing,* July–September, 2000, p. 60.

33. Carl Covalt Chambers. Oral history: "Interview with Carl Covalt Chambers" by Nancy Stern for the Charles Babbage Institute. Minneapolis: University of Minnesota, 1977, p. 16.

34. Ibid.

35. Ibid, p. 17.

36. Aspray, "Was Early Entry a Competitive Advantage?" p. 60.

37. Rheingold, *Tools for Thought,* p. 75

38. Edward Walterscheid. "The Need for a Uniform Government Patent Policy: The D.O.E. Example." *Harvard Journal of Law & Technology,* Vol. 3, Spring 1990, p. 111.

39. Ibid, p. 112.

40. Aspray, "Was Early Entry a Competitive Advantage?" p. 60.

41. Ibid., p. 61.

42. Rheingold, *Tools for Thought,* p. 70.

43. J. Presper Eckert. Oral history: "Interview with J. Presper Eckert, Chief Engineer, ENIAC Computer." by David Allison. Washington, DC:

National Museum of American History, Smithsonian Institution, 1988, p. 6.

44. Anne Fitzpatrick. "Teller's Technical Nemesis: The American Hydrogen Bomb and its Development with a Technological Infrastructure." *Techné: Journal of the Society for Philosophy and Technology.* Blacksburg: Virginia Tech, Vol. 3, No. 3, Spring 1998, p. 13.

45. Rheingold, *Tools for Thought,* p. 75.

46. Fitzpatrick, "Teller's Technical Nemesis," p.13.

47. Eckert, oral history, p.10.

48. Herbert R. J. Grosch. *Computer: Bit Slices from a Life.* Novato, CA: Third Millennium Books, 1991, p. 91.

49. Chambers, oral history, p. 27.

50. Grosch, *Computer,* pp. 73–74.

51. Eric Raymond. "The Cathedral and the Bazaar." *Matrix News,* May 1998, Vol. 8, No. 5.

52. Ibid., p. 6.

53. Maurice V. Wilkes. "A Tribute to Presper Eckert." *Communications of the ACM,* September 1995, Vol. 38, No. 9, p. 21.

Chapter Four

1. Raymond Ojserkis. *Beginnings of the Cold War Arms Race: The Truman Administration and the U.S. Arms Build-Up.* Westport, CT: Praeger, 2003, p. 6.

2. Ibid., p. 8.

3. Ibid., p. 8.

4. Ibid., p. 9.

5. Stephen Budiansky. "Closing the Book on Pearl Harbor." *Cryptologia.* April 2000, Vol. XXIV, No. 2. West Point, NY: United States Military Academy, pp. 119–120.

6. William Crowell. "Remembrances of Venona." <http://history.sandiego.edu / gen/text/coldwar/venona-crowell.html> (accessed November 02, 2009.

7. Samuel Snyder. "Influence of U.S. Cryptologic Organizations on the Digital Computer Industry." *Journal of Systems and Software,* 1 (1979), p. 7.

8. Mina Rees. "The Computing Program of the Office of Naval Research, 1946–1953." *Communications of the Association of Computing Machinery,* Vol. 30, No. 10, October, 1987, p. 832.

9. William Blanpied. "Science and Public Policy: The Steelman Report and the Politics of Post-World Ward II Science Policy." *Science and Technology Policy Yearbook: 1999.* Washington: American Association for the

Advancement of Science, 1999. <http://www.aaas.org/spp/yearbook / chap29.html>, p. 2. (accessed November 02, 2009).

10. Ibid., p. 3.

11. Kenneth Flamm. *Creating the Computer: Government, Industry, and High Technology.* Washington, D.C.: The Brookings Institute, 1988, p. 42.

12. Ibid., pp. 42–43.

13. Ibid., p. 35.

14. George Gray. "Engineering Research Associates and the Atlas Computer (UNIVAC 1101)." *Unisys History Newsletter*, Vol. 3, No. 3, June 1999, p. 2.

15. Flamm, *Creating the Computer*, p. 43.

16. Ibid., p. 44.

17. Gray, "UNIVAC 1101," p. 2.

18. Flamm, *Creating the Computer*, p. 45.

19. Ibid., p.45.

20. Ibid., p.45.

21. Richard Sprague. "A Western View of Computer History." *Communications of the ACM*, Vol. 15, No. 7, July 1972, p. 689.

22. Herman Wolk. "The Founding of the Air Force." *Air Force: Journal of the Air Force Association*, Vol. 79, No. 9, September 1996, p. 4.

23. Stuart Symington. Excerpt from remarks at swearing-in ceremony as first Secretary of the Air Force. Washington, D.C., September 18, 1947.

24. Ivan A. Getting. Oral history conducted by Frederik Nebeker, IEEE History Center, Rutgers University, New Brunswick, NJ, 1995, p. 4.

25. Stephen McFarland. "The Air Force in the Cold War, 1945–60: Birth of a New Defense Paradigm." *Airpower Journal*, Fall 1996, p. 7.

26. Ibid., p. 8.

27. Major Dik Daso. "Operation LUSTY: the U.S. Army Air Forces' Exploitation of the Luftwaffe's Secret Aeronautical Technology, 1944–45." *Aerospace Power Journal*, Spring 2002, p. 36.

28. Ibid., p. 40.

29. Nick Cook. *The Hunt for Zero Point*. New York: Broadway Books, 2002. pp. 65–66.

30. Guy Stever. *In War and Peace: My Life in Science and Technology.* Washington, D.C.: Joseph Henry Press, 2002, p. 308.

31. Ibid., p. 58.

32. Major Dik Daso. "Origins of Air Power: Hap Arnold's Early Career in Aviation Technology, 1903–1935." *Air Power Journal*, Winter 1996, p. 86.

33. Daso, "Operation LUSTY," p. 39.

34. Major Dik Daso. "Origins of Air Power: Hap Arnold's Command Years and Aviation Technology, 1936–1945." *Air and Space Power Journal.* Vol. XI, No. 3 (Fall 1997): Maxwell Air Force Base, Ala.: Air University Press. 1996, p. 95.

35. Theodore von Karman and Lee Edson. *The Wind and Beyond.* Boston: Little, Brown and Co., 1967, p. 243.

36. Stan Augarten. *Bit by Bit - An Illustrated History of Computers.* New York: Ticknor and Fields, 1984, pp. 138–139.

37. Marvin Wolf. "From Rocketeers to Solar Sailors." *Los Angeles Times Magazine*, November 14, 1999.

38. Daso, "Operation LUSTY," p. 37.

39. Daso, "Operation LUSTY," p. 30.

40. Guy Stever. Interview with Dik Daso, May 18, 1995.

41. Stever, *In War and Peace,* p. 58.

42. Daso, "Operation LUSTY," p. 32.

43. Arnold Pacey. *Technology in World Civilization.* Cambridge, MA: MIT Press, 1998, p. 204.

44. Stever, *In War and Peace,* p.59.

45. Hap Arnold letter to General Carl Spaatz, December 6, 1945.

46. Daso, "Operation LUSTY," p. 34.

47. Ibid., p.35

48. Arnold Pacey. *The Maze of Ingenuity.* Cambridge, MA: MIT Press, 1996, pp. 122–124.

49. Thomas Hughes. *American Genesis: A History of the American Genius for Invention.* New York: Penguin Group, 1990, p.1.

50. Daso, "Operation LUSTY," p.35.

51. Ibid.

52. William Aspray. "Was Early Entry a Competitive Advantage? U.S. Universities that Entered Computing in the 1940s." *IEEE Annals of the History of Computing,* July–September, 2000, pp. 44–45.

53. Larry Owens. "Where Are We Going, Phil Morse? Changing Agendas and the Rhetoric of Obviousness in the Transformation of Computing at MIT, 1939-1957." *IEEE Annals of the History of Computing,* Vol. 18, No. 4, 1996, p. 38.

54. J. R. Killian to Compton, "Memo on Coordinating our Activities in the Field of Analysis and Computation," October 4, 1945, AC4, Box 49, Folder 7, "Center for Analysis."

55. Jay Forrester. "Lincoln Laboratory, MIT Historical Comments." *Heritage Lecture Series.* November 26, 2001, p. 6.

56. Stuart Leslie. *The Cold War and American Science: The Military-Industrial-Academic Complex at MIT and Stanford.* New York: Columbia University Press, 1993, Chapter 1, p.9.

57. Ibid., Chapter 1, p. 4.

58. Thomas Hughes. *American Genesis: A History of the American Genius for Invention.* New York: Penguin Group, 1990, pp. 120–121.

59. Leslie, *Cold War and American Science,* Chapter 5, p. 2.

60. Owens, "Where Are We Going, Phil Morse?" p. 34.

61. Aspray, "Was Early Entry a Competitive Advantage?" p. 45.

62. Karl Wildes. *Electrical Engineering at the Massachusetts Institute of Technology.* Unpublished manuscript, Chapter 5, pp. 127–131.

63. Owens, "Where Are We Going, Phil Morse?" p. 34.

64. Leslie, *Cold War and American Science,* Chapter 1, p.6.

65. Ibid.

66. Gen. Dwight Eisenhower. "Memorandum for the Directors and Chiefs of the War Department General and Special Staff Divisions and Bureaus and the Commanding Generals of the Major Commands." April 30, 1946.

67. James Penick et al. *Politics of American Science: 1939–the Present.* Chicago: Rand McNally and Co., 1965, p. 100.

68. Louis Smullin. Interview with Stuart Leslie, Cambridge, Mass. January 20, 1988.

69. Leslie, *Cold War and American Science,* Chapter 1, p. 1.

70. Ibid., Flamm, p. 55.

71. Julius Stratton. "Memorandum to Members of the Research Laboratory of Electronics," May 1, 1949.

72. David J. Kelves. *The Physicists: The History of a Scientific Community in Modern America.* Cambridge, MA: Harvard University Press, 1995, p. 355.

73. Leslie, *Cold War and American Science,* Chapter 1, p. 7.

74. Conference with National Military Establishment, November 14–15, 1949. Washington, DC.

75. Leslie, *Cold War and American Science,* Chapter 5, p. 9.

76. Ibid., Chapter 1, p. 7.

77. Stever, *In War and Peace,* n. 35, p. 311.

78. Ibid., n. 45, p. 312.

79. Ibid., p. 83.

80. William T. Golden. Memorandum: Executive Office of the President. November 21, 1950.

81. Stever, *In War and Peace,* p. 94.

82. Getting, oral history, p. 5

83. McFarland, "The Air Force in the Cold War," p. 11.

84. Kent Redmond and Thomas Smith. *Project Whirlwind: A Case History in Contemporary Technology*. Bedford, MA: MITRE Corporation, 1975, Chapter 3, pp. 26–27.

Chapter Five

1. George Valley. "How the SAGE Development Began." *Annals of the History of Computing*, Vol. 7, No. 3, July 1985, p. 197.

2. Guy Stever. *In War and Peace: My Life In Science and Technology*. Washington, D.C.: Joseph Henry Press, 2002, p. 45.

3. Valley, "How the SAGE Development Began," p. 197.

4. Ibid., p. 198.

5. Robert Conquest. *The Great Terror: A Reassessment*. London: Oxford University Press, 1990.

6. Winston Churchill. *Address at MIT Mid-Century Convocation*. Boston Garden. March 31, 1949. Cambridge, MA: MIT News Service.

7. R. C. Williams. *Klaus Fuchs, Atom Spy*. Cambridge, MA: Harvard University Press, 1987, p. 18.

8. Michael I. Schwartz. "The Russian A-bomb: The Role of Espionage in the Soviet Atomic Bomb Project." *Journal of Undergraduate Science*, Summer 1996. Cambridge, MA: Harvard University, p. 105.

9. Ibid.

10. Ibid., p. 106.

11. Ibid., Reference Note 1, p.106.

12. Kenneth Flamm. *Creating the Computer: Government, Industry, and High Technology*. Washington, D.C.: The Brookings Institute, 1988, p. 55.

13. Kent Redmond and Thomas Smith. *Project Whirlwind: A Case History in Contemporary Technology*. Bedford, MA: MITRE Corporation, 1975, Chapter 4, p. 2.

14. Ibid., Chapter 4, pp. 11–12.

15. Ibid., Chapter 4, p. 21.

16. Ibid., Chapter 5, p. 7.

17. Robert Everett. "Whirlwind," in Nicholas Metropolis, J. Howlett, and Gian-Carlo Rota, eds., *A History of Computing in the Twentieth Century*. New York: Academic Press, 1980, p. 367.

18. Redmond and Smith. *Project Whirlwind*, Chapter 3, p. 18.

19. Kent Redmond and Thomas Smith. *From Whirlwind to MITRE: The R&D Story of the SAGE Air Defense Computer*. Cambridge, MA: MIT Press, 2000, p. 72.

20. Ibid.

21. Kent Redmond and Thomas Smith. "Lessons from Project Whirlwind." *IEEE Spectrum*, October 1977, p. 54.

22. Jay Forrester. "Lincoln Laboratory, MIT Historical Comments." *Heritage Lecture Series*. November 26, 2001, p. 6.

23. Robert Everett. Conversation with the author. October 2003.

24. D. H. Lawrence. "Terra Incognita." Selected Poetry of D. H. Lawrence. New York: Penguin Classics, 1989.

25. Perry Crawford. Master's thesis, "Automatic Control by Arithmetical Operations." Cambridge, MA: MIT. 1942, p.8.

26. David Israel. Master's thesis, "The Application of a High-Speed Digital Computer to the Present-Day Air Traffic Control System." Cambridge, MA: MIT. 1951, p.1.

27. Yaneer Bar-Yam. *Dynamics of Complex Systems*. Reading, MA: Addison-Wesley, 1997, p.12.

28. Redmond and Smith, *Project Whirlwind*, Chapter 3, p. 18.

29. Ibid.

30. Vannevar Bush. Interview with Henry Guerlac. Records of the Office of Strategic Research and Development (OSRD) Group 227. Boston: The National Archives, Northeast Region. August 20, 1944, pp. 5–6.

31. Redmond and Smith, *Project Whirlwind*, Chapter 4, p. 2.

32. Stuart Leslie. *The Cold War and American Science: The Military-Industrial-Academic Complex at MIT and Stanford*. New York: Columbia University Press, 1993, Chapter 1, p. 8.

33. Redmond and Smith, *Project Whirlwind*, Chapter 5, p. 16.

34. Ibid., Chapter 5, p. 13.

35. Forrester. *Heritage Lecture Series*. November 26, 2001, p. 6.

36. Redmond and Smith, *Project Whirlwind*, Chapter 4, p. 26.

37. J. W. Forrester and R. R. Everett. "Report L-1: Digital Computation for Anti-submarine Problem." Servomechanisms Laboratory report. Cambridge, MA: MIT, October 1, 1947; and "Report L-2: Information System of Interconnected Digital Computers." Servomechanisms Laboratory report. Cambridge, MA: MIT, October 15, 1947.

38. Redmond and Smith, *Project Whirlwind*, Chapter 4, p. 25.

39. Ibid.

40. Ibid.

41. Bruce Old. "Return on Investment in Basic Research Exploring a Methodology." Report to the Department of Naval Research, Department of the Navy. Bruce Old Associates, November 1981, p. 11.

42. Ibid.

43. Stever, *In War and Peace*, pp. 72–73.

44. Robert Seamans. Interview with Martin Collins. Oral History Project for the National Air and Space Museum, April 9, 1987, pp. 4–6.

45. Flamm, *Creating the Computer,* pp. 44–45.

46. Everett, "Whirlwind," p. 367.

47. Kevin Moore. *A Brief History of Aircraft Flight Simulation.*

48. http://homepage.ntlworld.com/bleep/SimHist1.html, 2005, pp. 3–4.

49. Ibid, pp. 6–7.

50. Old, "Return on Investment," pp. 12–13.

51. Jay Forrester. Section, "Reliability in Digital Computers," from Report L-21, "Analysis of Digital Computer Laboratory Proposed 1950–1951 Budget," by Jay Forrester et al. 1951, pp. 3–4.

52. Old, "Return on Investment," p. 12.

53. Jay Forrester. "Whirlwind's Success." Transcript of talk from symposium: *Whirlwind's Genesis and Descendants.* Boston: Computer Museum, October 18, 1987, p. 3.

54. Jay Forrester. Oral History with David Allison for the National Museum of American History, March 24, 1998, p. 5.

55. Redmond and Smith, *From Whirlwind to MITRE,* p. 74.

56. Mina Rees. "The Computing Program of the Office of Naval Research, 1946–1953." *Communications of the Association of Computing Machinery,* Vol. 30, No. 10, October 1987, p. 831.

57. Old, "Return on Investment," p. 23.

58. Rees, "The Computing Program," p. 838.

59. Richard Sprague. "A Western View of Computer History." *Communications of the Association of Computing Machinery,* Vol. 15, No. 7, 1972, p. 687.

60. Bijker, Hughes, and Pinch. *The Social Construction of Technology.* Cambridge, MA: MIT Press, 1989, pp. 10–50.

61. Valley, "How the SAGE Development Began," p. 218.

62. Ibid.

63. Larry Owens. "Where Are We Going, Phil Morse? Changing Agendas and the Rhetoric of Obviousness in the Transformation of Computing at MIT, 1939–1957." *IEEE Annals of the History of Computing,* Vol. 18, No. 4, 1996, p. 35.

64. James Birkenstock. "Pioneering: On the Frontier of Electronic Data Processing, a Personal Memoir." *IEEE Annals of the History of Computing,* January–March 2000, p. 19.

65. Nancy Stern. *From ENIAC to UNIVAC: Appraisal of the Eckert-Mauchly Computers.* Boston: Digital Press, 1981, p. 111.

66. Karl L. Wildes and Nilo A. Lindgren. *A Century of Electrical Engineering and Computer Science at MIT, 1882–1982*. Cambridge, MA: MIT Press, 1985, p. 233.

67. Owens, "Where Are We Going, Phil Morse?" p. 38.

68. Letter from S. Caldwell to V. Bush, September 25, 1945. Library of Congress, Bush Papers, Box 18.

69. Wildes and Lindgren, pp. 232–233.

70. Owens, "Where Are We Going, Phil Morse?" p. 34.

71. Ibid, p. 37.

72. Karl L. Wildes. Unpublished manuscript, "Electrical Engineering at the Massachusetts Institute of Technology," Chapter 5, pp. 127–131.

73. David Alan Grier. "The Human Computer and the Birth of the Information Age." The Joseph Henry Lecture. Washington, DC: Philosophical Society. May 11, 2001.

74. Roland Omnès. *Quantum Philosophy: Understanding and Interpreting Contemporary Science*. New Jersey: Princeton University Press, 1999, pp. 141–146.

75. Ibid, p. 200.

76. Chris Bissell. "A Great Disappearing Act: the Electronic Analog Computer." IEEE Conference on the History of Electronics, Bletchley Park: UK, June 28–30, 2004, p. 3.

77. Redmond and Smith, *Project Whirlwind*, Chapter 5, p. 16.

78. Ibid., Chapter 10, p. 14.

79. Ibid., Chapter 5, p. 16.

80. John Hutchins. "From First Conception to First Demonstration: The Nascent Years of Machine Translation, 1947–1954, A Chronology." *Machine Translation*, Vol. 12, No. 3, p. 195.

81. Norbert Wiener. *Cybernetics, or Control and Communication in the Animal and Machine*, Cambridge, MA: MIT Press, 1961, p. 21.

82. Ibid, p. 15

83. Hutchins, "From First Conception," p. 247.

84. Owens, "Where Are We Going, Phil Morse?" p. 38.

85. Yehoshua Bar-Hillel. "Mechanical Translation: Needs and Possibilities." Cambridge, MA: MIT Research Laboratory of Electronics, p. 6.

86. Hutchins, "From First Conception," p. 198.

87. Ibid.

88. Ibid., p. 247.

89. Rees, "The Computing Program," p. 832.

90. Hutchins, "From First Conception," p. 197.

91. Rees, "The Computing Program," p. 832.

92. Warren Weaver. "Science and Complexity." *American Scientist*, Vol 36, 1948, p. 539.

93. Ibid.

94. Larry Owens. "Vannevar Bush and the Differential Analyzer: The Text and Context of an Early Computer." *Technology and Culture*, Vol. 27,

95. No. 1, pp. 63–95, January 1986.

96. John Hutchins. "First Steps in Mechanical Translation." Norwich, UK: University of East Anglia, 1997, p.1.

97. Redmond and Smith, *Project Whirlwind*, Chapter 9, p. 8.

98. Ibid., Chapter 5, p. 23.

99. Ibid., Chapter 8, p. 3.

100. Ibid., Chapter 5, p. 16.

101. Ibid., Chapter 7, p. 20.

102. Ibid., Chapter 5, pp. 40–42.

103. Rees, "The Computing Program," p. 838.

104. Harry Huskey. "The National Bureau of Standards Western Automatic Computer (SWAC)." *Annals of the History of Computing*, Vol. 2, No. 2, (April 1980), pp. 111–121.

105. Hutchins. "First Steps," p. 4.

106. Redmond and Smith, *Project Whirlwind*, Chapter 7, p. 20.

107. Ibid., Chapter 7, p. 23.

108. Ibid., Chapter 7, p. 15.

109. Ibid., Chapter 6, p. 28 and Chapter 7, p. 23.

110. Ibid, Chapter 6, p. 30.

111. Ibid, Chapter 7, p. 27.

112. Jay W. Forrester, H. R. Boyd, R. R. Everett, H. Fahnestock, R. A. Nelson. "Forecast for Military Systems using Electronic Digital Computers," Memorandum L-3, Servomechanisms Laboratory, Cambridge, MA: MIT, 1948.

113. Redmond and Smith, *Project Whirlwind*, Chapter 7, p. 26.

114. Ibid, Chapter 7, p. 32.

115. Ibid, Chapter 7, p. 26.

116. Kenneth Flamm. *Creating the Computer: Government, Industry, and High Technology.* Washington, D.C.: Brookings Institute, 1988, p. 55.

117. David Dickson. *The New Politics of Science.* Chicago: University of Chicago Press, 1988, p. 119.

118. Rees, "The Computing Program," p. 831.

119. Flamm, *Creating the Computer,* p.55.

A. Kent Redmond and Thomas Smith

Bright Boys is indebted to two other bright boys who came before it: Kent Redmond and Thomas Smith. If Whirlwind could have had its own Boswell, Messrs Redmond and Smith would have been that chronicler. For over three decades, maybe longer, they researched and wrote voluminously on this most singular machine and its very singular builders. Their devotion to their subject and the energy of their scholarship over the years enriched the literature on the technology of early computing. They rolled their research out into the scholarly and professional literature in scores of articles and papers, as well as in a thousand pages or more in these three volumes:

Project Whirlwind: A Case History in Contemporary Technology (Manuscript Edition), 1975.

Project Whirlwind: The History of a Pioneer Computer, 1980.

From Whirlwind to MITRE: The R&D Story of the SAGE Air Defense Computer, 2000.

To this writer, their sections of notes and bibliographies are breathtaking. There can't be a library, library shelf, book, journal, notebook, serial publication, box of papers, folder of memorabilia, private letter, public document, note, fax, phone call or e-mail that their hands and eyes missed in compiling research to tell their Whirlwind story. Although *Bright Boys* covers different ground for a different audience, their research, which in many instances is the only research extant on the subject, helped me to put my book together. The trail of Information Technology is impossible to follow without their help; and they saved this writer a ton of shoe leather along the way. The boys and I are grateful for their guidance.

B. A Brief History of the Electron Tube

The electron tube or vacuum tube (American English), known in England as a thermionic valve or more simply as a "valve" was popular from 1920 to 1960. Used in most electronic devices, vacuum tubes now serve only in specialized applications. For most purposes, the vacuum tube has been replaced by the much smaller and less expensive transistor, either as a discrete device or in an integrated circuit.

Vacuum tubes, or thermionic valves, are arrangements of electrodes surrounded by an airtight, temperature-resistant glass bulb. The tubes are designed to plug into tube sockets for easy replacement. There is a filament inside each vacuum tube that when it gets hot releases electrons into the vacuum or airtight space within the bulb. That process is called thermionic emission. Thermionic emission, once known as the "Edison Effect" (after Thomas Edison who first observed the emission), is the flow of a cloud of negatively charged electrons. These electrons are drawn to a metal plate inside the bulb that is positively charged, called the anode. This results in a current of electrons flowing to the heated plate. This type of vacuum tube is called a diode (Greek "di" meaning two).

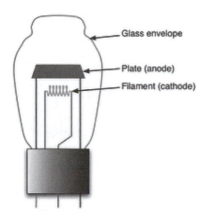

Fleming's Diode.

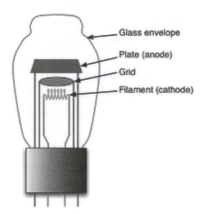

De Forest's Diode.

John Ambrose Fleming, scientific adviser to the Marconi Company (England), invented the diode in 1904 based on his observation of the Edison Effect. Fleming's invention of the vacuum tube or thermionic valve heralded the beginning of modern electronics. It enabled wireless and later electronics technology to move forward. Lee De Forest (US) placed another electrode, a bent wire or screen called a grid, between the filament and plate in 1906. De Forest discovered that the current flow from filament to plate depended on the voltage applied to the grid, and that the current drawn by the grid was very low, being composed of the electrons that are intercepted by the grid.

The resulting three-electrode device was an excellent and very sensitive amplifier of voltages. De Forest called his invention the "audion," but it is better known as a triode: the vacuum tube equivalent of a transistor. Further improvement came with the invention of a two-grid tube called a "tetrode," meaning four active electrodes, followed by a three-grid tube called a "pentode," meaning five electrodes. Electron or vacuum tubes were used by the thousands in early computers as on/off switches for computer operations. Today, such tubes are replaced by millions of miniature on/off switches on microprocessor chips in a modern computer.

Tubes were everywhere in the early generations of electronic devices, such as radios, televisions, and early computers such as the Colossus, which used 2,000 tubes; and ENIAC, which used nearly 18,000 tubes. Unlike transistors, vacuum tubes are inherently immune to the electromagnetic pulse effect of nuclear explosions. This property kept them in use for military applications long after transistors had replaced them elsewhere. Vacuum tubes are still used for specialized audio amplifiers, notably for electric guitar amplification, and for very high-powered applications such as microwave ovens, industrial radio-frequency heating, and power amplification for broadcasting.

Many thanks for the research information on "Electron Tubes" above courtesy of Wikipedia, WordIQ, and the Free Software Foundation.

Chapter Six

1. Jay Forrester. "Conversation with Jay Forrester." Transcript from audiotape interview by Christopher Evans. Computer History Project. Washington, DC: Smithsonian Institution, 1975, p. 5.

2. Emerson Pugh. "Ferrite Core Memories that Shaped an Industry." *IEEE Transactions on Magnetics*, Vol. 5, September 1984, p. 1499.

3. "Report of the Secretary of Commerce." United States Department of Commerce, Washington DC: GPO, 1946, pp. xxvi–xxvii.

4. "Report of the Secretary of Commerce." pp. 17–29.

5. James Birkenstock. "Pioneering: On the Frontier of Electronic Data Processing, a Personal Memoir." *IEEE Annals of the History of Computing*, January–March 2000, p. 20.

6. Forrester. "Conversation with Jay Forrester," p.5.

7. Forrester. "Conversation with Jay Forrester," p.5.

8. Kent Redmond and Thomas Smith. *Project Whirlwind: A Case History in Contemporary Technology*. Bedford, MA: MITRE Corporation, 1975, Chapter 9, p. 10.

9. Redmond and Smith. *Project Whirlwind,* Chapter 9, p. 11.

10. Redmond and Smith. *Project Whirlwind,* Chapter 9, p. 2.

11. Redmond and Smith. *Project Whirlwind,* Chapter 9, p. 3.

12. Stephen Johnson. *The United States Air Force and the Culture of Innovation: 1945–1965*. Washington, D.C.: Air Force History and Museums Program, 2002, p. 122.

13. Redmond and Smith. *Project Whirlwind,* Chapter 9, p. 5.

14. Redmond and Smith. *From Whirlwind to MITRE: The R&D Story of the SAGE Air Defense Computer*. Cambridge, MA: MIT Press, 2000, pp. 62–63.

15. Redmond and Smith. *From Whirlwind to MITRE,* p. 14.

16. Redmond and Smith. *From Whirlwind to MITRE,* p. 18.

17. Redmond and Smith. *From Whirlwind to MITRE,* p. 17.

18. Jay Forrester. "Notebook Supplement 50JWF2." *MIT Computation Book No. 50*, May 4, 1949.

19. Jay Forrester. "Notebook Supplement 49JWF15." *MIT Computation Book No. 49*, January 17, 1949, p.15.

20. Forrester. "Conversation with Jay Forrester," p. 5.

21. Pehr Anderson et al. "The Core of Engineering Revolutions." *Core Report*. Cambridge, MA: Laboratory of Computer Science, 2004, p. 4.

22. George Valley. "How the SAGE Development Began." *Annals of the History of Computing*, Vol. 7, No. 3, July 1985, p. 216.

23. Eric Weiss. "An Wang, 1920–1990." *IEEE Annals of the History of Computing,*Vol. 15, No. 1, 1993, p. 62.

24. Ibid., p. 62.

25. Ibid., p. 62

26. Pugh, p. 1500.

27. Weiss, p.63.

28. Valley, p. 199.

29. Johnson, p. 123.

30. Valley, p. 199.

31. Kenneth Shaffel. *The Emerging Shield: The Air Force and the Evolution of Continental Air Defense, 1945–1960.* Washington, D.C.: Office of Air Force History, United States Air Force, 1991, p. 83.

32. Shaffel, p. 83.

33. Shaffel, p. 79.

34. Anthony Cave Brown. *The Last Hero: Wild Bill Donovan.* New York: Times Books, p. 802.

35. Shaffel, p. 71.

36. Shaffel, p. 87.

37. Herman Wolk. "The Founding of the Air Force." *Air Force: Journal of the Air Force Association* Vol. 79, No. 9, September 1996, p.4.

38. Muir Fairchild (Maj). "Foreign Developments in Anti-Aircraft Defense." (Lecture at ACTS), November 8, 1939.

39. Shaffel, p. 86.

40. Gordon Saville. "Interim Program for Aircraft Control and Warning in the Continental U.S., Alaska, Presentation to Secretary of Defense Forrestal." September 9, 1948.

41. Redmond and Smith. *Project Whirlwind,* Chap. 9, pp. 10–11.

42. Raymond Ojserkis. *Beginnings of the Cold War Arms Race: The Truman Administration and the U.S. Arms Build-Up.* Westport, CT: Praeger, 2003, p. ix.

43. Valley, p. 197.

44. Ibid., p. 197.

45. Ibid., p. 199.

46. Ibid. p. 199.

47. "Report on Electronic Digital Computers by the Consultants to the Chairman of the Research and Development Board." Washington: Research and Development Board, June 15, 1950, p. 9.

48. Jay Forrester. "Statement of the Status of Project Whirlwind Prepared for the Research and Development Board," L-24, Servomechanism Laboratory, MIT, May 10, 1950.

49. Shaffel, p. 14.

50. Ibid., p. 63.

51. Francis Ankenbrandt (Maj. Gen.). "Opening Remarks at Presentation of Aircraft Control and Warning System for Alaska and the United States," November 19, 1947.

52. Shaffel, p. 70.

53. President Harry S. Truman, memo to James Forrestal, June 3, 1948.

54. Shaffel, p. 91.

55. John Piotrowski (General). *Strategic Synchronization: The Relationship between Strategic Offense and Defense."* Washington, D.C.: The Heritage Foundation, 2002, pp. 2–3.

56. Shaffel, pp. 91–92.

57. Hoyt Vandenberg (General). Personal journal, August 24, 1948.

58. Shaffel, p. 98.

59. Ibid., p. 99.

60. Ibid., p. 101.

61. Guy Stever. *In War and Peace: My Life In Science and Technology.* Washington, D.C.: Joseph Henry Press, 2002, pp. 98–99.

62. Stever, p. 99.

63. Muir Fairchild (General). Memo to General Vandenberg. December 8, 1949.

64. Valley, p. 199.

65. David Winkler. *Searching the Skies: The Legacy of the United States Cold War Defense Radar Program.* Langley, VA: USAF Headquarters Air Combat Command, 1997, p. 20.

66. John V. Harrington. "Radar Data Transmission." *Annals of the History of Computing*, Vol. 5, No. 4, October 1983, p. 371.

67. Ibid.

68. Ibid., p. 372.

69. Ibid.

70. Ibid., p. 206.

71. Ibid., p. 207.

72. Ibid.

73. Ibid.

74. Ibid.

75. Charles Seife. *Decoding the Universe.* New York: The Penguin Group, 2006, pp. 10, 76.

76. Valley, p. 198.

77. Howard Rheingold. *Tools for Thought.* Cambridge, MA: MIT Press, 2000, p. 112.

78. Ibid., p. 114.

79. Valley, p. 207.

80. Redmond and Smith. *Project Whirlwind,* Chapter 10, p. 11.

81. Valley, p. 208.

82. Jay Forrester. *Computation Book No. 49.* Entry for January 27, 1950, p. 83.

83. Pehr Anderson et al. "The Core of Engineering Revolutions." *Core Report.* Cambridge, MA: Laboratory of Computer Science, 2004, p. 31.

84. Memorandum L-18, February 15, 1950, p.1.

85. Valley, p. 208.

86. Ibid., p. 209.

87. Ibid.

Chapter Seven

1. David Winkler. *Searching the Skies: The Legacy of the United States Cold War Defense Radar Program.* Langley, VA: USAF Headquarters Air Combat Command, 1997, p. 21.

2. Les Earnest. *Internet Creation Myths.* <http://www.stanford.edu/ ~learnest/ net.htm>, July 2004, p.2. (accessed November 02, 2009).

3. Steven Poole. "Computing and the Net: Switched On." *The Guardian,* April 12, 2003, p.4.

4. Jim Ray. "SAGE: Cold War Forerunner to the Information Age." <http:// www.eskimo.com/~wow-ray/sage28.html > p.1 (accessed November 02, 2009).

5. Carl Posey. "The Thin Aluminum Line." *Air & Space Magazine* (Smithsonian). December/January 2007, p. 56.

6. Albert Wohlstetter and Henry Rowen. *Objectives of the United States Military Posture.* RAND, RM-2373. Santa Monica, CA: RAND Corporation, May 1, 1959, p. 26.

7. Richard F. McMullen. "History of Air Defense Weapons." *ADC Historical Study No. 14.* Washington: Historical Division, Office of Information. Headquarters U.S.A.F., 1968, p.18.

8. Poole, p. 4.

9. Les Earnest. "Vigilance and Vacuum Tubes: The SAGE System 1956–1963." Speech transcript: Computer Museum History Center, Mountain View, CA, May 19, 1998, p.16.

10. Federation of American Scientists and John Schance, System Program Director. "Strategic Automated Command Control System." <http://www.fas.org/nuke/guide/usa/c3i/saccs.htm> January 1999.

11. Ibid.

12. Paul Edwards. "Vigilance and Vacuum Tubes: The SAGE System 1956–1963.".Speech transcript: Computer Museum History Center, Mountain View, CA, May 19, 1998, p.10.

13. Glenn E. Burgos. "Programming the American Aerospace Industry, 1954–1964." *Business and Economic History*, Vol. 22, No. 1, Fall 1993, p. 215.

14. Claude Baum. *The System Builders: The Story of SDC*. Santa Monica, CA: SDC, 1981, p. 43.

15. John F. Jacobs. "SAGE Overview." *IEEE Annals of the History of Computing*, Vol. 5, No. 4, October–December 1983, p. 324.

16. Ronald Kline. "What is Information Theory a Theory Of?" *Proceedings of the History and Heritage of Scientific and Technological Information Systems*. W. Boyd Rayward and Mary Ellen Bowden, editors. Philadelphia, PA: American Society of Information Science and Technology, 2004, p. 17.

17. W. A. Hosier. "Pitfalls and Safeguards in Real-Time Digital Systems." *IRE Transactions on Engineering Management,* June 1961, p. 312.

18. Ray, p.3.

19. *Time*. "Search for Safety." May 27, 1957, p.23.

20. Paul Edwards. *The Closed World: Computers and the Politics of Discourse in Cold War America*. Cambridge, MA: MIT Press, 1996, p. 101.

21. James Birkenstock. "Pioneering: On the Frontier of Electronic Data Processing, a Personal Memoir." *IEEE Annals of the History of Computing*, January–March 2000, p.26.

22. Thomas Haigh. "A Veritable Bucket of Facts: Origins of the Data Base Management System." SIGMOD Record, Vol. 35., No. 2, June 2006, p. 33.

23. Robert F. Martina. "Letter to Bunny White: Archivist at Lucent Technologies." Murray Hill, NJ: Lucent Technologies. October 15, 1997, p. 2.

24. Ibid.

25. Les Earnest. *Stories about SAGE*. Ed Thelen (editor), p.7. Online at: http://ed-thelen.org/sage-1.html (accessed November 02, 2009)

26. C. Robert Wieser. "From World War II Radar Systems to SAGE." Speech at the Computer Museum. Boston, MA: *Computer Museum Report*, Vol. 22, Spring 1988, p. 5.

27. Ibid.

28. Mike Loewen. *The AN/FSQ-7 on TV and in the Movies*. May 2, 2006. <http://ripsaw.cac.psu.edu/~mloewen/Q7>

29. George Dyson. *Darwin Among the Machines*. Presentation at the Reality Club, New York, July 8, 1997, p. 6.

30. Mina Rees. "The Computing Program of the Office of Naval Research, 1946–1953." *Communications of the Association of Computing Machinery*, Vol. 30, No. 10, October 1987, p. 824.

31. Whirlwind 1. Product booklet. Cambridge, MA: MIT Servomechanisms Laboratory, August 15, 1951, p. 10.

32. Ibid., p. 16.

33. Stephen Johnson. *The United States Air Force and the Culture of Innovation: 1945–1965*. Washington, D.C.: Air Force History and Museums Program, 2002, p.131.

34. Ibid.

35. Peter Norton. *Inside the IBM PC*. Bowie, MD: R. J. Brady Co., 1983, p. 5.

36. Kent Redmond and Thomas Smith. *Project Whirlwind: A Case History in Contemporary Technology*. Bedford, MA: MITRE Corporation, 1975, Chapter 9, p. 17.

37. George Valley. "How the SAGE Development Began." *Annals of the History of Computing*, Vol. 7, No. 3, July 1985, p. 208.

38. Redmond and Smith. *Project Whirlwind*. Chapter 9, p. 19.

39. Valley, see Note 10, p. 210.

40. Redmond and Smith. *Project Whirlwind*. Chapter 9, p. 40.

41. Valley, p. 210.

42. Thomas A. Sturm. *The USAF Scientific Advisory Board: Its First Twenty Years, 1944–1964*. Washington, DC: USAF Historical Division Liaison Office, 1967, p. 39.

43. Valley, p. 199.

44. Ibid.

45. Ibid., p. 203.

46. Ibid., p. 205.

47. David Mindell. "Bodies, Ideas, and Dynamics: Historical Perspectives on Systems Thinking in Engineering." (working paper) Cambridge, MA: MIT Engineering Systems Division, 2002, p. 13.

48. Valley, p. 204.

49. Wieser, p.2.

50. Earnest, *Internet Creation Myths*, p.2.

51. Thomas Marill and Lawrence Roberts. "Toward a Cooperative Network of Time-Shared Computers." Fall Conference, American Federation of Information Processing Societies, November 7–10, 1966. San Francisco, California.

52. Stephen B. Johnson. *The United States Air Force and the Culture of Innovation: 1945–1965*. Washington, D.C.: Air Force History and Museums Program, 2002, p.26.

53. Valley, p.204.

54. Ibid.

55. Ibid., p. 210.

56. Ibid., p.211.

57. William Aspray. "Was Early Entry a Competitive Advantage? U.S. Universities that Entered Computing in the 1940s." *IEEE Annals of the History of Computing*, July–September 2000, p. 42.

58. *The Digital Computer Laboratory of the Massachusetts Institute of Technology*. Report R-199-1. October 1, 1953, p. 2.

59. Ibid., p. 3.

60. Ibid., p. 2.

61. Saul Rosen. "Electronic Computers: A Historical Survey." *Computing Surveys*, Vol. 1, No. 1, March 1969.

62. Roger Mills. Personal reflections from paper to be delivered at the *Last Hurrah Conference* (conference cancelled), <http:// www. palosverdes.com/last-hurrah> 2004, p.1. (accessed November 02, 2009).

63. Ibid.

64. Rosen, p. 10.

65. Nancy Stern. "The Eckert-Mauchly Computers." *Technology and Culture*, Vol. 23, No. 4, October 1982, p. 581.

66. Birkenstock, p.17.

67. Birkenstock, p. 18.

68. Herbert Grosch. *Computer: Bit Slices from a Life*. Novato, CA: Third Millennium Books, 1991, p. 117.

69. Stern, p. 581.

70. Birkenstock, p. 19.

71. Leslie Goff. "UNIVAC Predicts Winner of 1952 Election." Computerworld Flashback CNN.com, April 30, 1999.

72. Birkenstock, p.18.

73. Werner Buchholz. Interview. "Computer Oral History Collection, 1969–1973." Washington, DC: National Museum of American History, July 20, 1973, p. 8.

74. Nat Rochester. Interview. "Computer Oral History Collection, 1969–1973." Washington, DC: National Museum of American History, July 24, 1973, p. 1.

75. Morton Astrahan and John Jacobs. "History of the Design of the SAGE Computer—The AN/FSQ-7." *Annals of the History of Computing*, Vol. 5, No. 4, October 1983, p. 344.

76. Stuart Leslie. *The Cold War and American Science: The Military-Industrial-Academic Complex at MIT and Stanford.* New York: Columbia University Press, 1993, p. 185.

77. Rosen, p. 20.

78. George Gray. "Engineering Research Associates and the Atlas Computer." *Unisys History Newsletter*, Vol. 3, No. 3, June 1999, p. 1.

79. James Boone. "The WWII Cryptologic Heritage of the United States' Computer and Communications Industries." Fort Meade, MD: National Cryptologic Museum Foundation, Inc., 2004, p. 3.

80. Ibid., p. 4.

81. Gray, p. 2.

82. George Gray. "The UNIVAC 1102, 1103, and 1104." *Unisys History Newsletter*, Vol. 6, No. 1, January 2002, p. 1.

83. Drew Pearson. "Washington Merry-Go-Round." *The Washington Post*. August 16, 1950, p.12.

84. Ibid.

85. Gray, "Engineering Research Associates . . .", p. 5.

86. Rosen, "Computing Surveys . . .", Note 7, p. 35.

87. Saul Rosen. "Recollections of the Philco Transac S-2000." *Annals of the History of Computing*, April–June 2004, p. 35.

88. Ibid., p. 21.

89. Ibid., p. 35.

90. Ibid.

91. Rosen, "Computing Surveys . . .", p. 17.

92. *Elecom 120* product brochure, Underwood Computer Co., 1953, p. 2.

93. Richard Sprague. "A Western View of Computer History." *Communications of the Association of Computing Machinery*, Vol. 15, No. 7, 1972, p. 689.

94. Rosen, "Computing Surveys . . .", p. 19.

95. Ibid., p. 11.

Chapter Eight

1. Emerson Pugh. "Ferrite Core Memories that Shaped an Industry." *IEEE Transactions on Magnetics*, Vol. 5, September 1984, p. 1501.

2. Saul Rosen. "Electronic Computers: A Historical Survey." *Computing Surveys*, Vol. 1, No. 1, March 1969, p. 14.

3. Ibid.

4. Pugh, p. 1501.

5.   Redmond and Smith. *From Whirlwind to MITRE: The R&D Story of the SAGE Air Defense Computer*. Cambridge, MA: MIT Press, 2000, p. 437.

6.   Stan Augarten. *Bit by Bit—An Illustrated History of Computers*. New York: Ticknor and Fields, 1984, p. 208.

7.   Thomas J. Watson Jr. *Father, Son & Co.: My Life at IBM and Beyond*. New York: Bantam Books, 1990, p. 233.

8.   Jay Forrester. Memorandum to A. G. Hill. November 5, 1952 and May 12, 1953.

9.   A. P. Kromer. "Minutes of Joint MIT-IBM Conference held in Hartford, Connecticut, January 20, 1953." January 26, 1953.

10. Morton Astrahan and John Jacobs. "History of the Design of the SAGE Computer—The AN/FSQ-7." *Annals of the History of Computing*, Vol. 5, No. 4, October 1983, pp. 345–346.

11. Rosen, p. 9.

12. Astrahan and Jacobs, p. 346.

13. Redmond and Smith, p. 47.

14. Ibid.

15. Raymond Ojserkis. *Beginnings of the Cold War Arms Race: The Truman Administration and the U.S. Arms Build-Up*. Praeger: Westport, CT, 2003, p. 1

16. Ibid.

17. Harry Truman. Interview with Carleton Kent, *Washington Sun-Times*, January 16, 1953.

18. Charles E. Bohlen. *Witness to History, 1929–1969*, New York: W. W. Norton, 1973, p. 303.

19. Yoneyuki Sugita. *Pitfall or Panacea: The Irony of U.S. Power in Occupied Japan 1945–1952*. London: Routledge, 2003, pp. 99–100.

20. *Time*, March 26, 1956.

21. *Time*, December 10, 1956.

22. James Birkenstock. "Pioneering: On the Frontier of Electronic Data Processing, a Personal Memoir." *IEEE Annals of the History of Computing*, January–March 2000, pp. 27–28.

23. Ibid.

24. Ibid., p.29.

25. Ibid.

26. Ibid., p. 28.

27. Augarten, p. 210.

28. Guy Stever. *In War and Peace: My Life In Science and Technology*. Washington, D.C.: Joseph Henry Press, 2002, p. 314.

29. *Whirlwind I*. Brochure for Electronic Computer Division, Servo-mechanisms Laboratory. August 15, 1951, p. 14.

30. George Valley. "How the SAGE Development Began." *Annals of the History of Computing*, Vol. 7, No. 3, July 1985, p. 212.

31. Vannevar Bush. *Science, the Endless Frontier*. Washington: National Science Foundation, 1945, p 5.

32. Stever, p. 96.

33. Thomas Sturm. The USAF Scientific Advisory Board: Its First Twenty Years, 1944–1964. Washington, DC: USAF Historical Division Office, 1967, p. 35.

34. Stever, p. 94.

35. Ivan Getting. *All in a Lifetime: Science in the Defense of Democracy*. New York: Vantage Press, 1989, p. 350.

36. Valley, p. 212.

37. Jay Stratton to James Killian, October 20, 1950, MIT Archives.

38. Louis Ridenour. "Memorandum for General Saville, Subject: Proposed Augmentation of ADSEC Activities." November 20, 1950.

39. Valley, p. 211.

40. Ibid., p. 212.

41. Ibid.

42. Getting, pp. 231–232.

43. Stuart Leslie. *The Cold War and American Science: The Military-Industrial-Academic Complex at MIT and Stanford*. New York: Columbia University Press, 1993, p. 10.

44. Valley, p. 213.

45. Ibid., p. 212.

46. General Hoyt Vandenberg. "Letter to Dr. James R. Killian." December 15, 1950.

47. Valley, p. 213.

48. Redmond and Smith, pp. 92–93.

49. Valley, p. 213.

50. Ibid.

51. Leslie, p. 21.

52. Redmond and Smith, p. 216.

53. Ken Olsen. "An Interview with Ken Olsen." Transcript of an Oral History Interview by David Allison. Washington, DC: Smithsonian Institution, September 28–29, 1988.

54. Ibid.

55. Gordon Bell, Craig Mudge, and John McNamara. *Computer Engineering*. Maynard, MA: Digital Press, 1978, p. 124.

56. Olsen.

57. William Buxton. "Interaction at Lincoln Laboratory in the 1960's: Looking Forward—Looking Back." *CHI 2005 Panels*, April 2–7, 2005, p. 1163.

58. Adrian Johnstone. "The Pre-History of the Digital Equipment Corporation." *Computer Resurrection*, Issue 4, pp. 15–16.

59. Eldon Hall. "From the Farm to Pioneering with Digital Computers: An Autobiography." *Annals of the History of Computing*, April–June 2000, p. 24.

60. Larry Owens. "Where Are We Going, Phil Morse? Changing Agendas and the Rhetoric of Obviousness in the Transformation of Computing at MIT, 1939–1957." *Annals of the History of Computing*, Vol. 18, No. 4, 1996, p. 35.

61. Hall, p. 25.

62. Ibid.

63. Hall, p. 26.

64. Enders Robinson. Interview by Andrew Goldstein, IEEE History Center, New Brunswick, NJ, March 6, 1997.

65. Ibid.

66. Robert Dean Clark. "Biography: Enders Robinson." *The Leading Edge*, February 9, 2006, p. 1.

67. Office of Naval Research Mathematical Science Division. *Digital Computer*, Vol. 5, No. 2, April 1953, p. 1.

68. Robinson.

69. Office of Naval Research, *Digital Computer*, p. 2.

70. *Making Electrons Count.* "Solving a Problem on M.I.T.'s Electronic Digital Computer: Whirlwind I." Movie. Cambridge, MA: MIT's Digital Computer Laboratory, 1953.

71. Douglas Ross. "A Personal View of the Personal Workstation: Some Firsts in the Fifties." *Annals of the History of Computing*, January 1986, p. 21.

72. *A Conversation with Doug Engelbart.* Documentary film script by Frode Hegland and Fleur Klijnsma. London, 2003.

73. Ross, p. 19.

74. Jay Forrester. "Digital Information Storage in Three Dimensions Using Magnetic Cores." *Journal of Applied Physics*, Vol. 22, No. 1, January 1951.

75. Valley, p. 219.

76. Olsen, p. 21.

77. Ibid.

78. Ibid.

79. Forrester, "Digital Information Storage . . . ", p. 46.

80. Jennifer J. Armiger. "Magnetic-Core Memory Hits 50-Year Mark." *IEEE Engineer Online*, August 2003.

81. Redmond and Smith, p. 315.

82. Ross, p. 24.

83. Jay Forrester. Email to the author, September 30, 2009.

84. Valley, p. 217.

85. Ibid.

86. Ross, p. 22.

87. Redmond and Smith, pp. 256–257.

88. Valley, p. 218.

89. C. Robert Wieser. "From World War II Radar Systems to SAGE." Speech at the Computer Museum. Boston, MA: *Computer Museum Report*, Vol. 22, Spring 1988, p.4.

90. Redmond and Smith, pp. 268–269.

91. C. Robert Wieser. "Cape Cod System." *Annals of the History of Computing*, Vol. 7, No. 3, July 1985, p. 368.

92. Interview with Paul Rosen by Michael Geselowitz. New Brunswick, NJ: IEEE History Center, April 22, 2004, p. 16.

93. Wieser, p. 369.

Chapter Nine

1. Jay Forrester. Transcript: "Video History Interview with Jay Forrester." by David Allison, Washington, DC: National Museum of American History, March 24, 1998, p. 8.

2. Ibid.

3. *Time*, "Outstanding Scientists," June 7, 1954. <http://www.time.com/time/magazine/article/0,9171,806881,00.html> (accessed November 02, 2009).

4. C. Robert Wieser. "From World War II Radar Systems to SAGE." Speech at the Computer Museum. Boston, MA: *Computer Museum Report*, Vol. 22, Spring 1988, p. 5.

5. William Radford. "MIT: Lincoln Laboratory: Its Origin and First Decade." *Technology Review*, January 1962, pp. 15–18.

6. Carl Overhage. "Notes on Interactions between MIT and Lincoln Laboratory." February 14, 1963, AC 134, p.19.

7. Carl Overhage. "Lincoln Laboratory and the Greater MIT Community." *Technology Review*, January 1962, p. 24.

8. *An Interview with Wesley Clark*. Transcript of audio interview conducted by Judy O'Neill. Minneapolis, MN: Charles Babbage Institute. May 3, 1990.

9. John Naughton. *A Brief History of the Future: From Radio Days to Internet Years in a Lifetime.* New York: Overlook Press, 2000, p. 71.

10. Fernando Corbato and Robert Fano. "Time-sharing on Computers," chapter in *Information—A Scientific American Book*, San Francisco, CA: 1966, p. 76

11. Naughton, *A Brief History of the Future,* p. 81.

12. William Buxton. "Interaction at Lincoln Laboratory in the 1960's: Looking Forward—Looking Back." *CHI 2005 Panels*, April 2–7, 2005, p. 1165.

13. James Wong. "Vigilance and Vacuum Tubes: The SAGE System 1956–1963." Speech transcript: Computer Museum History Center, Mountain View, CA, May 19, 1998, p. 10.

14. Ibid., p. 11.

15. Ibid., p. 12.

16. Ibid., p. 13.

17. Ibid., p. 14.

18. Ibid.

19. Ibid.

20. *Business Week*, February 29, 1964, pp. 70–90.

21. Jules Schwartz. "The Development of JOVIAL." Association for Computing Machinery SIGPLAN Notices, Vol. 13, No. 8, August 1978, p. 209.

22. Stan Augarten. *Bit by Bit—An Illustrated History of Computers.* New York: Ticknor and Fields, 1984, p. 215.

23. John Backus. "Programming in America in the 1950s—Some Personal Impressions," in Nicholas Metropolis, J. Howlett, and Gian-Carlo Rota, eds., *A History of Computing in the Twentieth Century.* New York: Academic Press, 1980, p. 130.

24. Ibid.

25. Ibid., pp. 130–131.

26. Ibid., pp. 125–126.

27. John Correll. "How the Air Force got the ICBM." *Air Force Magazine*, July 2005, p. 73.

28. Guy Stever. *In War and Peace: My Life In Science and Technology.* Washington, D.C.: Joseph Henry Press, 2002, p. 129.

29. Ibid., p. 125.

30. Stephen Johnson. "The Organizational Roots of American Economic Competitiveness in High Technology." Paper presented at the Conference of R&D Investment and Economic Growth in the 20th Century. Haas Business School. University of California, Berkeley, 1999, p. 5.

31. Stephen Johnson. *The United States Air Force and the Culture of Innovation: 1945–1965.* Washington, D.C.: Air Force History and Museums Program, 2002, p. 172

32. Johnson, "The Organizational Roots," p. 19.

33. Henry Tropp et al., "A Perspective on SAGE: Discussion." *Annals of the History of Computing,* Vol. 5, No. 4, October 1983, p. 394

34. Johnson, "The Organizational Roots,"pp. 19–20.

35. David Mindell. "Bodies, Ideas, and Dynamics: Historical Perspectives on Systems Thinking in Engineering." (working paper) Cambridge, MA: MIT Engineering Systems Division, 2002, p.17.

36. Kent Redmond and Thomas Smith. *From Whirlwind to MITRE: The R&D Story of the SAGE Air Defense Computer.* Cambridge, MA: MIT Press, 2000, p. 110.

37. Tropp et al., "A Perspective on SAGE," p. 389.

38. John Pike. "Strategic Air Command." Global Security. http://www.globalsecurity.org/wmd/agency/sac.htm, June 20, 2005.

39. Carlos Cuadra. "Online before the Internet, Part 3: Early Pioneers Tell Their Stories: Carlos Cuadra," Vol. 11, No. 9, October 2003, p. 2.

40. Robert Futrell. *Ideas, Concepts, Doctrine: Basic Thinking in the United States Air Force 1907–1960.* Maxwell AFB, Alabama: Air University Press, 1989, p. 512.

41. Carl Posey. "The Thin Aluminum Line." *Air & Space Magazine* (Smithsonian). December/January 2007, p. 55.

42. Curtis LeMay. News Release 609-58, Office of Public Information, Department of Defense (ESDHO), 1958.

43. Curtis LeMay. *The Command Post.* Roslyn AFB, New York, July 1958, p.4.

44. Stever, *In War and Peace,* p. 133.

45. Ibid., 112.

46. Ibid., p. 133.

47. Howard Rheingold. *Tools for Thought: The History and Future of Mind-Expanding Technology.* Cambridge, MA: MIT Press, 2000, p. 144.

48. John Naughton. *A Brief History of the Future: From Radio Days to Internet Years in a Lifetime.* New York: Overlook Press, 2000, pp. 78–79.

49. Rheingold, *Tools for Thought,* p. 145.

50. Ibid., p. 149.

51. "Automation in the Office." National Office Management Association, Willow Grove, PA, 1957, p.19.

52. Rheingold, *Tools for Thought,* p. 149.

53. Thomas Haigh. "Lost in Translation: Total Systems from War Room to Board Room, 1954–1968." Paper presented at the Society of History of Technology annual meeting: *The Social Architectures of Cold War Electronics.* Toronto: November 2002, p. 12.

54. Ibid., p. 14.

55. Harold Leavitt and Thomas Whistler. "Management in the 1980s." *Harvard Business Review*, Vol. 36, November–December 1958, pp. 41–48.

56. Haigh, "Lost in Translation," p. 13.

57. Andrew Leonard. "Do-it-yourself giant brains!" Salon Technology. http://www.salon.com, June 22, 2000.

58. Rheingold, *Tools for Thought,* pp. 178–179.

59. Ibid., p. 180.

60. Haigh, "Lost in Translation," p. 19.

61. Ibid., p. 20.

62. Information Processing Techniques Office. "Internet History." http://www.livinginternet.com/i/ii.htm, September 10, 2004.

Bibliography

Allvin, David W. (Lt. Colonel USAF). "Airpower Strategy in the Interwar Years: Not Ready for Prime Time." Master's thesis, National Defense University, National War College, Maxwell AFB, Alabama, GA, 2003.

Anderson, Pehr et al. "The Core of Engineering Revolutions." Core Report, MIT Laboratory of Computer Science, Cambridge, MA, 2004.

Ankenbrandt, Francis (Maj. General) "Opening Remarks at Presentation of Aircraft Control and Warning System for Alaska and the United States." Washington, D.C., November 19, 1947.

Armiger, Jennifer J. "Your Engineering Heritage: Magnetic-Core Memory Hits 50-Year Mark." *IEEE-USA Today's Engineer Online*, August 2003. (Available at http://www.todaysengineer. org/ 2003/Aug/history.asp.)

Arnold, Henry H. (General). "Air Force in the Atomic Age." In *One World or None: A Report to the Public on the Full Meaning of the Atomic Bomb*, Dexter Masters and Katherine Way (eds.), pp. 26–32. New York: McGraw-Hill, 1946.

Aspray, William. "Was Early Entry a Competitive Advantage? U.S. Universities That Entered Computing in the 1940s." *IEEE Annals of the History of Computing*, Vol. 22, No. 3, July 2000, pp. 60–75.

Astrahan, Morton and Jacobs, John. "History of the Design of the SAGE Computer—The AN/FSQ-7." *Annals of the History of Computing*, Vol. 5, No. 4, October 1983, pp. 344–352.

Augarten, Stan. *Bit by Bit: An Illustrated History of Computers.* New York: Ticknor & Fields, 1984.

Backus, John. "Programming in America in the 1950s—Some Personal Impressions." In *A History of Computing in the Twentieth Century,* Nicholas Metropolis, J. Howlett, and Gian-Carlo Rota (eds.), pp. 130–137. New York: Academic Press, 1980.

Baugher, Joseph. "Soviet B-29, a Tupolev Tu-4 Story," *Russian Aviation FAQ.* http://aeroweb.lucia.it/rap/RAFAQ/Tu-4.html, July 17, 1997.

Baum, Claude. *The System Builders: The Story of SDC.* Santa Monica, CA: SDC Press, 1981.

Beasley, J.E. "OR-Notes." Course notes for Operations Research and Management Science, Department of Mathematical Sciences, Brunel University, West London, UK, 2004.

Bell, Gordon, Mudge, Craig and McNamara, John. *Computer Engineering.* Maynard, MA: Digital Press, 1978.

Bennett, Stuart. "A Brief History of Automatic Control." *IEEE Control Systems,* Vol. 16, No. 3, June 1996, pp. 20–23.

Bijker, Wiebe, Hughes, Thomas, and Pinch, Trevor. *The Social Construction of Technology.* Cambridge, MA: MIT Press, 1989.

Birkenstock, James W. "Pioneering: On the Frontier of Electronic Data Processing, a Personal Memoir." *IEEE Annals of the History of Computing,* Vol. 22, No. 1, January–March 2000, pp. 19–51.

Blanpied, William A. "Science and Public Policy: The Steelman Report and the Politics of Post-World Ward II Science Policy." Chapter 29 of *AAAS Science and Technology Policy Yearbook: 1999.* Washington, D.C.: American Association for the Advancement of Science, 1999.

Bohlen, Charles E. *Witness to History, 1929–1969.* New York: W. W. Norton, 1973.

Boone, James. *The WWII Cryptologic Heritage of the United States' Computer and Communications Industries.* Fort Meade, MD: National Cryptologic Museum Foundation, Inc., 2004.

Bracken, Paul "Warning and Intelligence." Chapter 2 of *The Command and Control of Nuclear Forces*. New Haven, CT: Yale University Press, 1983.

Bragg, Steve. "The SCR-584: Radar Tribute Page." http://www.hamhud.net/darts/scr584.html, 2007.

Brown, Anthony Cave. *The Last Hero: Wild Bill Donovan*. New York: Times Books, 1983.

Buchholz, Werner. "Interview with Werner Buchholz." Interview in Computer Oral History Collection, National Museum of American History, Washington, D.C., July 20, 1973.

Budiansky, Stephen. "Closing the Book on Pearl Harbor." *Cryptologia*, Vol. XXIV, No. 2, April 2000, pp. 119–120.

Burgos, Glenn F. "Programming the American Aerospace Industry, 1954–1964." *Business and Economic History*, Vol. 22, No. 1, Fall 1993, pp. 215–224.

Bush, Vannevar. *Science, the Endless Frontier*. Washington, D.C.: The National Science Foundation, 1945.

Buxton, William. "Interaction at Lincoln Laboratory in the 1960's: Looking Forward—Looking Back." In *CHI'05 Extended Abstracts on Human Factors in Computing Systems*, pp. 1162–1167. New York: ACM Press, 2005.

Byrd, Martha. *Kenneth Walker: Airpower's Untempered Crusader*. Maxwell AFB, Alabama, GA: Air University Press, 1997.

Caldwell, Samuel. "Letter from S. Caldwell to V. Bush." September 25, 1945. Library of Congress, Bush Papers, Box 18.

Capra, Fritjof. "The Turning Point: A Science of Living Systems." http://www.sfsu.edu/~ihhs/class/382/1.doc, 1982.

Capra, Fritjof. *The Web of Life: A New Synthesis of Mind and Matter*. London: Flamingo, 1997.

Carpenter, Joel. "UFOs & Nuclear Weapons," *National Investigations Committee on Aerial Phenomena*. http://www.nicap.org/ncp/nn-020628.htm, June 2002.

Ceruzzi, Paul E. "A View from 20 Years as a Historian of Computing." *IEEE Annals of the History of Computing*, Vol. 23, No. 4, October–December 2001, pp. 22–27.

Ceruzzi, Paul. "Electronic Calculators." Chapter 7 of *Computers Before Computing*, William Aspray (ed.). Ames, IA: Iowa State University Press, 1990.

Chambers, Charles Covalt. "Interview with Carl Chambers." Charles Babbage Institute, University of Minnesota, Minneapolis, MN, 1977.

Clark, Robert Dean. "Biography: Enders Robinson." *The Leading Edge*, Vol. 4, No. 2, February 1985, pp. 16–20.

Clark, Wesley. "An Interview with Wesley Clark." Interview by Judy O'Neill, Charles Babbage Institute, University of Minnesota, Minneapolis, MN, May 3, 1990.

Clark, Wesley. Email to Tom Green. October 13, 2009.

Cook, Nick. *The Hunt for Zero Point.* New York: Broadway Books, 2002.

Corbato, Fernando and Fano, Robert. "Time-sharing on Computers." In *Information—A Scientific American Book*, pp. 76–83. San Francisco, CA: W. H. Freeman, 1966.

Correll, John. "How the Air Force got the ICBM." *Air Force Magazine*, Vol. 88, No. 7, July 2005, pp. 73–77.

Corum, Dr. James S. "Airpower Thought in Continental Europe between the Wars." In *Paths of Heaven: The Evolution of Air Power Theory*, Philip Meilinger (Colonel) (ed.), PP. 151–181. Alabama, GA: Air University Press, 1997.

Craven, W. F. and J. L. Cate. "The Air Corps Prepares for War, 1939–41." Chapter 4 of *The Army Air Forces in World War II*, Volume I "Plans and Early Operations: January 1939 to August 1942," Wesley Frank and Cate Craven James Lea (eds.), pp. 112–134. Chicago: The University Press of Chicago, 1950.

Crowell, William. "Remembrances of Venona." http://history.acusd.edu/gen/text/coldwar/venona-crowell.html, no date.

Cuadra, Carlos. "Online before the Internet, Part 3: Early Pioneers Tell Their Stories: Carlos Cuadra." *Searcher*, Vol. 11, No. 9, October 2003, pp. 2–7.

Daso, Dik Alan. "Operation LUSTY: The U.S. Army Forces' Exploitation of the Luftwaffe's Secret Aeronautical Technology, 1944–45." *Aerospace Power Journal*, Vol. XVI, No. 1, Spring 2002, pp. 36–42.

Daso, Dik Alan. "Origins of Air Power: Hap Arnold's Command Years and Aviation Technology, 1936–1945." *Air and Space Power Journal*, Vol. XI, No. 3, Fall 1997, pp. 94–113.

Daso, Dik Alan. "Origins of Air Power: Hap Arnold's Early Career in Aviation Technology, 1903–1935." *Airpower Journal*, Vol. X, No. 4, Winter 1996, pp. 86–101.

Delaney, William P. and Ward, William W. "Radar Development at Lincoln Laboratory: An Overview of the First Fifty Years." *Lincoln Laboratory Journal*, Vol. 12, No. 2, 2000, pp. 147–166.

Digital Computer Laboratory of MIT. *Making Electrons Count: Solving a Problem on M.I.T.'s Electronic Digital Computer 'Whirlwind I.'* Movie. Cambridge, MA: Digital Computer Laboratory of MIT, 1953.

Digital Computer Laboratory of MIT. "Report R-199-1." Massachusetts Institute of Technology, Cambridge, MA, October 1, 1953.

Doolittle, James Harold. "Reminiscences of James Harold Doolittle." Interview by Kenneth. W. Leish. Oral History Research Office, Columbia University, New York, April, 1960.

Dyson, George. "Darwin Among the Machines." Presentation at the Reality Club, New York, NY, July 8, 1997.

Earnest, Les. "Internet Creation Myths." http://www.stanford.edu/~learnest/net.htm, July 1, 2004.

Earnest, Les. "Vigilance and Vacuum Tubes: The SAGE System 1956–1963." Speech given at Computer Museum History Center, Mountain View, CA, May 19, 1998.

Earnest, Les. Various contributions to *Stories about SAGE,* Ed Thelen (ed.). http://yarchive.net/risks/sage_bomarc.html, February 20, 1999.

Eckert, Presper J. "Interview with J. Presper Eckert, Chief Engineer, ENIAC Computer." Interview by National Museum of American History, Smithsonian Institution, Washington, D.C., 1988.

Edwards, Paul N. "The World in a Machine: Origins and Impacts of Early Computerized Global Systems Models." In *Systems, Experts, and Computers,* Agatha Hughes and Thomas Hughes (eds.), pp. 221–235. Cambridge, MA: MIT Press, 1997.

Edwards, Paul N. *The Closed World: Computers and the Politics of Discourse in Cold War America.* Cambridge, MA: MIT Press, 1996.

Eisenhower, Dwight David (General). "Memorandum for the Directors and Chiefs of the War Department General and Special Staff Divisions and Bureaus and the Commanding Generals of the Major Commands," April 30, 1946.

Electronic Computer Division. *Whirlwind I* brochure. MIT Servomechanisms Laboratory, Cambridge, MA, August 15, 1951.

"ENIAC." *New York Times,* February 16, 1946, p. 1.

McGraw-Hill staff. "The SCR-268 Radar." *Electronics.* September 1945, pp. 100–104.

Everett, Robert. "Whirlwind." In *A History of Computing in the Twentieth Century,* Nicholas Metropolis, J. Howlett, and Gian-Carlo Rota (eds.), pp. 384–392. New York: Academic Press, 1980.

Everett, R. R. and Swain, F. E. "Whirlwind I Computer Block Diagrams," Report R-127, two volumes, Servomechanisms Laboratory, Massachusetts Institute of Technology, Cambridge, MA, September 4, 1947.

Faber, Peter R. (Lt. Col. USAF). "Interwar U.S. Army Aviation and the Air Corps Tactical School: Incubators of American Air Power." Chapter 6 of *The Paths of Heaven: The Evolution of Air Power Theory,* Phillp S. Meilinger (Colonel, USAF) (ed.). Maxwell AFB, Alabama, GA: Air University Press, 1997.

Fairchild, Muir (General). Memo to General Vandenberg. December 8, 1949.

Fairchild, Muir (Major). "Foreign Developments in Anti-aircraft Defense." Lecture at the Air Corps Tactical School (ACTS), Maxwell Field, Montgomery, Alabama, November 8, 1939.

Federation of American Scientists and John Schance, System Program Director. "Strategic Automated Command and Control Systems." http://www.fas.org/nuke/guide/usa/c3i/saccs.htm, January 1999.

Fielding, Raymond. *The March of Time, 1935–1951*. New York: Oxford University Press, 1978.

Finney, Robert T. *History of the Air Corps Tactical School, 1920–1940*. Maxwell AFB, Alabama, GA: Research Studies Institute, Air University, 1955.

Fitzpatrick, Anne. "Teller's Technical Nemesis: The American Hydrogen Bomb and its Development within a Technological Infrastructure." *Techné: Journal of the Society for Philosophy and Technology*, Vol. 3, No. 3, Spring 1998, pp. 13–17.

Flamm, Kenneth. *Creating the Computer: Government, Industry, and High Technology*. Washington, D.C.: The Brookings Institute, 1988.

Forrester, Jay W. "Lincoln Laboratory, MIT: Historical Comments." Heritage Lecture Series lecture given on the 50th Anniversary of Lincoln Laboratory, Lexington, MA, November 26, 2001.

Forrester, Jay W. "Digital Information Storage in Three Dimensions Using Magnetic Cores." *Journal of Applied Physics*, Vol. 22, No. 1, January 1951, pp. 46–52.

Forrester, Jay W. "Memorandum to A.G. Hill. November 5, 1952 and May 12, 1953." Lincoln Steering Committee minutes and Report L-95, January, 1953.

Forrester, Jay W. "Notebook Supplement 49JWF15." *MIT Computation Book No. 49*, January 17, 1949,

Forrester, Jay W. "Notebook Supplement 50JWF2." *MIT Computation Book No. 50*, May 4, 1949.

Forrester, Jay W. "Statement of the Status of Project Whirlwind Prepared for the Research and Development Board." Report L-24, Servomechanism Laboratory, MIT, Cambridge, MA, May 10, 1950.

Forrester, Jay W. "The Beginnings of System Dynamics." Banquet talk for the System Dynamics Society, Stuttgart, Germany, July 13, 1989.

Forrester, Jay W. Transcript of "Video Interview with Jay Forrester." Interview by David Allison, National Museum of American History, Washington, D.C., March 24, 1998.

Forrester, Jay W. "Whirlwind's Success." Talk presented at the symposium Whirlwind's Genesis and Descendants Computer Museum, Boston, MA, October 18, 1987.

Forrester, Jay W. Email to Tom Green, September 30, 2009.

Forrester, Jay W. Personal letter to Eric A. Weiss, June 5, 1992.

Forrester, Jay W. Personal letter to Eric A. Weiss, October 14, 1992.

Forrester, Jay W., Boyd, W. H. R., Everett, R. R., Fahnestock, H., and Nelson, R. A. "Forecast for Military Systems using Electronic Digital Computers." Memorandum L-3, Servomechanisms Laboratory, Massachusetts Institute of Technology, Cambridge, MA, 1948.

Evans, Christopher. "Conversation with Jay Forrester." Adapted from *Pioneers of Computing*, Tape 4. London: Science Museum, 1975.

Futrell, Robert. *Ideas, Concepts, Doctrine: Basic Thinking in the United States Air Force 1907–1960*. Maxwell AFB, Alabama, GA: Air University Press, 1989.

Getting, Ivan "Interview with Ivan Getting." Interview by Frederik Nebeker, IEEE History Center, Rutgers University, New Brunswick, NJ, 1995.

Getting, Ivan. *All in a Lifetime: Science in the Defense of Democracy*. New York: Vantage Press, 1989.

Goebel, Greg. "Fortress in Development: Model 299 through B-17E." http://www.faqs.org/docs/air/avb171.html, 2002.

Goebel, Greg. "Radar and the Battle of Britain." http://www.vectorsite.net/ttwiz_02.html, 2002.

Goff, Leslie. "UNIVAC predicts winner of 1952 election." *CNN.com*. http://www.cnn.com/TECH/computing/9904/30/1952.idg/, April 30, 1999.

Golden, William T. Memorandum to Executive Office of the President, Washington, D.C., November 21, 1950.

Goldstine, Herman. "Computers at the University of Pennsylvania's Moore School, 1943–1946." *Proceedings of the American Philosophical Society*, Vol. 136, No. 1, 1992, pp. 62–77.

Gray, George. "Engineering Research Associates and the Atlas Computer (UNIVAC 1101)." *Unisys History Newsletter.* Vo. 3, No. 3, June 1999, pp. 1–3.

Gray, George. "The UNIVAC 1102, 1103, and 1104." *Unisys History Newsletter,* Vol. 6, No. 1, January 2002, pp. 2–4.

Grier, David Alan. "The Human Computer and the Birth of the Information Age." Joseph Henry Lecture of the Philosophical Society of Washington, Washington, D.C., May 11, 2001.

Grosch, Herbert. *Computer: Bit Slices from a Life.* Novato, CA: Third Millennium Books, 1991.

Haigh, Thomas. "Lost in Translation: Total Systems from War Room to Board Room, 1954–1968." Paper presented at the Society of History of Technology Annual Meeting, Toronto, Canada, November 10–14, 2002.

Haigh, Thomas. "A Veritable Bucket of Facts: Origins of the Data Base Management System." *SIGMOD Record,* Vol. 35, No. 2, June 2006, pp. 31–37.

Haigh, Thomas. "Inventing Information Systems: The Systems Men and the Computer, 1950–1968." *Business History Review,* Vol. 75, Spring 2001, pp. 15–61.

Hall, Eldon. "From the Farm to Pioneering with Digital Control Computers: An Autobiography." *IEEE Annals of the History of Computing,* Vol. 22, No. 2, April–June, 2000, pp. 20–27.

Hallion, Richard P. "Control of the Air: The Enduring Requirement." Special Studies, Air Force History and Museums Program, Bolling Air Force Base, Washington, D.C., September 8, 1999.

Halstead-Nussloch, Richard. "Computers and Society: Professional Practices and Ethics: Lecture 6—Five Times One Thousand." http://tapestry.spsu.edu/drrich/classes/cs2642/eniac6.htm, 2001.

Harrington, John V. "Radar Data Transmission." *Annals of the History of Computing,* Vol. 5, No. 4, October 1983, pp. 370–375.

Harris, Lynden T. (Bucky). "The DEW Line Chronicles." http://www.lswilson.ca/dewhist-a.htm, n.d.

Hegland, Frode and Klijnsma, Fleur. *A Conversation with Doug Englebart*. Documentary film. London: The Invisible Revolution, 2005.

Hosier, W.A. "Pitfalls and Safeguards in Real-Time Digital Systems with Emphasis on Programming." *IRE Transactions on Engineering Management*, Vol. EM-8, No. 2, June 1961, pp.99–115.

Hughes, Thomas. *American Genesis: A History of the American Genius for Invention*. New York: Penguin Group, 1990.

Hughes, Thomas. *Rescuing Prometheus*. New York: Pantheon, 1998.

Jacobs, John F. "SAGE Overview." *IEEE Annals of the History of Computing, Vol. 5*, No. 4, October–December 1983, pp. 321–344.

Johnson, Stephen B. "The Organizational Roots of American Economic Competitiveness in High Technology." Paper presented at the Conference of R&D Investment and Economic Growth in the 20th Century, Haas Business School, University of California, Berkeley, CA, March 28, 1999.

Johnson, Stephen B. *The United States Air Force and the Culture of Innovation: 1945–1965*. Washington, D.C.: Air Force History and Museums Program, 2002.

Johnstone, Adrian. "The Pre-history of the Digital Equipment Corporation." *Computer Resurrection*, Issue 4, Summer 1992, pp. 15–16.

Kaufman, Allen. "Assembling America's Private Arsenal for Democracy, 1920–1961." *Business and Economic History*, Vol. 26, No. 1, 1997, pp. 252–265.

Kelves, David J. *The Physicists: The History of a Scientific Community in Modern America*. Cambridge, MA: Harvard University Press, 1995.

Kent, Carleton. Interview of Harry Truman. *Washington Sun-Times*, January 16, 1953, p. 6.

Kline, Ronald. "What is Information Theory a Theory Of?" In *Proceedings of the History and Heritage of Scientific and Technological Information Systems*, W. Boyd Rayward and Mary Ellen Bowden (eds.), pp. 17-22. Philadelphia, PA: American Society of Information Science.

Kromer, A. P. "Minutes of Joint MIT-IBM Conference held in Hartford, Connecticut, January 20, 1953." Memorandum M-1810, Digital Computer Laboratory, Cambridge, MA, January 26, 1953.

Kuhn, Thomas S. *The Structure of Scientific Revolutions.* Chicago: The Chicago University Press,1996.

Leavitt, Harold and Whistler, Thomas. "Management in the 1980s." *Harvard Business Review*, Vol. 36, November–December 1958, pp. 41–48.

Leonard, Andrew. "Do-it-yourself giant brains!" *Salon Technology.* http://www.salon.com/technology/fsp/2000/06/22/chapter_2_part_two/print.html, June 22, 2000.

Leslie, Stuart. *The Cold War and American Science: The Military-Industrial-Academic Complex at MIT and Stanford.* New York: Columbia University Press, 1993.

Licklider, J.C.R. "Man-Computer Symbiosis." *IRE Transactions on the Human Factors in Electronics,* Vol. HFE-1, No. 1, March 1960, pp. 4–19.

Loewen, Mike. "The AN/FSQ-7 on TV and in the Movies." http://ripsaw.cac.psu.edu/~mloewen/Q7,May 2, 2006.

MacKenzie, Donald. *Inventing Accuracy.* Cambridge, MA: MIT Press, 1993.

Mahoney, Michael S. "The History of Computing in the History of Technology." *Annals of the History of Computing,* Vol. 10, No. 2, 1988, pp. 113–122.

Marill, Thomas and Roberts, Lawrence. "Toward a Cooperative Network of Time-Shared Computers." Paper presented at Fall Conference, American Federation of Information Processing Societies, San Francisco, CA, November 7–10, 1966.

Martina, Robert F. "Letter to Bunny White: Archivist at Lucent Technologies." October 15, 1997. (Available at http://www.smecc.org/sage_a_n_fsq-7.htm.)

Mathematical Science Division, Office of Naval Research. "Whirlwind Notes." *Digital Computer*, Vol. 5, No. 2, April 1953, p. 1.

Maurer, Maurer. *Aviation in the U.S. Army: 1919–1939.* Washington, D.C.: Office of Air Force History, 1987.

Mayo, Wayland. "The Russian B-29 Clone: The TU-4 Story." http://www.rb-29
.net/HTML/03RelatedStories/03.03shortstories/03.03.10contss.htm, accessed
November 7, 2009.

McFarland, Stephen. "The Air Force in the Cold War, 1945–60: Birth of a New
Defense Paradigm." *Airpower Journal*, Vol. X, No. 3, Fall 1996, pp. 7–13.

McMullen, Richard F. *History of Air Defense Weapons,* ADC Historical Study No.
14. Washington, D.C.: Office of Information, Historical Division, Headquarters
USAF, 1968.

Melville, Herman. *Moby Dick*. New York: Modern Library, 1992.

Mills, Roger. "Last Hurrah Didn't Make It." http://www.palosverdes.com/lasthurrah,
2004.

Mindell, David A. "Bodies, Ideas, and Dynamics: Historical Perspectives on
Systems Thinking in Engineering." Working paper, MIT Engineering Systems
Division, Cambridge, MA, 2002.

Mindell, David, Segal, Jerome, and Gerovitch, Slava. "From Communications
Engineering to Communications Science: Cybernetics and Information
Theory in the United States, France, and the Soviet Union." In *Science and
Ideology: A Comparative History,* Mark Walker (ed.), pp. 66-95. New York:
Routledge, 2003.

Moeller, Stephen P. (Colonel). "Vigilant and Invincible." *Air Defense Artillery,*
May–June 1995, pp. 4–14.

Nema National Office Management Committee. "Automation in the Office." Survey
Summary 21, National Office Management Association, Willow Grove, PA,
1957.

Naughton, John. *A Brief History of the Future: From Radio Days to Internet Years in
a Lifetime.* New York: Overlook Press, 2000.

Nisan, Noam and Schocken, Shimon. *The Elements of Computing Systems.*
Cambridge, MA: MIT Press, 2005.

Ojserkis, Raymond P. *Beginnings of the Cold War Arms Race: The Truman
Administration and the U.S. Arms Build-up.* Westport, CT: Praeger, 2003.

Olsen, Ken. "An Interview with Ken Olsen." Interview by David Allison, Smithsonian Institution, Washington, D.C., September 28–29, 1988.

Omnès, Roland. *Quantum Philosophy: Understanding and Interpreting Contemporary Science.* Princeton, NJ: Princeton University Press, 1999.

Overhage, Carl. "Lincoln Laboratory and the Greater MIT Community." *Technology Review*, January 1962, pp. 24–27.

Overhage, Carl. "Notes on Interactions between MIT and Lincoln Laboratory." Technical report AC 134, Lincoln Laboratory, Lexington, MA, February 14, 1963.

Owens, Larry. "Where Are We Going, Phil Morse? Changing Agendas and the Rhetoric of Obviousness in the Transformation of Computing at MIT, 1939–1957." *IEEE Annals of the History of Computing*, Vol. 18, No. 4, 1996, pp. 32–37.

Pacey, Arnold. *Technology in World Civilization.* Cambridge, MA: MIT Press, 1998.

Pacey, Arnold. *The Maze of Ingenuity.* Cambridge, MA: MIT Press, 1996.

Pearson, David E. *The World Wide Military Command and Control System.* Maxwell AFB, Alabama, GA: Air University Press, 2000.

Pearson, Drew. "Washington Merry-Go-Round" Syndicated article, August 16, 1950. (Typescript available from American University Library Special Collections, http://dspace.wrlc.org/doc/get/2041/23130/b09f20-0816zdisplay .pdf.)Penick, James et al. The *Politics of American Science: 1939 to the Present.* Chicago: Rand McNally & Co., 1965.

Perry, James D. "Air Corps Experimentation in the Interwar Years: A Case Study." *Joint Force Quarterly*, No. 22, Summer 1999, pp. 40–49.

Pike, John. "Strategic Air Command." Global Security. http://www.globalsecurity .org/wmd/agency/sac.htm, June 20, 2005.

Piotrowski, General John L. *Strategic Synchronization: The Relationship between Strategic Offense and Defense.*" Washington, D.C.: The Heritage Foundation, 2002.

Poole, Steven. "Switched On." *The Guardian*, April 12, 2003, p. 4.

Posey, Carl. "The Thin Aluminum Line." *Air & Space Magazine,* January 1, 2007, p. 56. (Available at http://www.airspacemag.com/history-of-flight/16045877. html.)

Pugh, Emerson W. and Aspray, William. "Creating the Computer Industry." *IEEE Annals of the History of Computing,* Vol. 18, No. 2, 1996, pp. 7–17.

Pugh, Emerson. "Ferrite Core Memories that Shaped an Industry." *IEEE Transactions on Magnetics,* Vol. 5, No. 5, September 1984, pp. 1501–1506.

Radford, William. "MIT: Lincoln Laboratory: Its Origin and First Decade." *Technology Review,* VOL, NO, January 1962, pp. 15–18.

Ramo, Simon. "Interview with Simon Ramo." Interview by Martin Collins. Oral History Project, National Air and Space Museum, Washington, D.C., June 27, 1988.

Ray, Jim "SAGE: Cold War Forerunner to the Information Age." http://www .eskimo.com/~wow-ray/sage28.html, 2006.

Redmond, Kent and Smith, Thomas. *Project Whirlwind: A Case History in Contemporary Technology.* Bedford, MA: MITRE Corporation, 1975.

Redmond, Kent and Smith, Thomas. "Lessons from Project Whirlwind." *IEEE Spectrum,* Vol. 14, No. 10, October 1977, pp. 26–27.

Redmond, Kent and Smith, Thomas. *From Whirlwind to MITRE: The R&D Story of the SAGE Air Defense Computer.* Cambridge, MA: MIT Press, 2000.

Rees, Mina. "The Computing Program of the Office of Naval Research, 1946–1953." *Communications of the ACM,* Vol. 30, No. 10, October 1987, pp. 829–837.

Research and Development Board. "Report on Electronic Digital Computers by the Consultants to the Chairman of the Research and Development Board." Department of Defense, Washington, D.C., June 15, 1950.

Rheingold, Howard. *Tools for Thought: The History and Future of Mind-Expanding Technology.* Cambridge, MA: MIT Press, 2000.

Ridenour, Louis. "Memorandum for General Saville, Subject: Proposed Augmentation of ADSEC Activities." November 20, 1950.

Robinson, Enders. "Interview with Enders Robinson." Interview by Andrew Goldstein, IEEE History Center, New Brunswick, NJ, March 6, 1997.

Rochester, Nathaniel. "Interview with Nate Rochester." Interview in Computer Oral History Collection, Oral History Project, National Museum of American History, Washington, D.C., July 24, 1973.

Rosen, Paul. "Interview with Paul Rosen." Interview by Michael Geselowitz, IEEE History Center New Brunswick, NJ, April 22, 2004.

Rosen, Saul. "Electronic Computers: A Historical Survey." *ACM Computing Surveys*, Vol. 1 No. 1, March 1969, pp. 10–33.

Rosen, Saul. "Recollections of the Philco Transac S-2000." *Annals of the History of Computing*, Vol. 26, No. 2, April–June 2004, pp. 30–34.

Ross, Douglas. "A Personal View of the Personal Work Station: Some Firsts in the Fifties." In *Proceedings of the ACM Conference on the History of Personal Workstations*, pp. 19–48. New York: ACM Press, 1986.

Ross, Stewart Halsey. *Strategic Bombing by the United States in World War II: The Myths and the Facts.* Jefferson, NC: McFarland & Company, Inc., 2003.

Saville, Gordon (General). "Interim Program for Aircraft Control and Warning in the Continental U.S., Alaska, Presentation to Secretary of Defense Forrestal." Briefing, Washington, D.C., September 9, 1948.

Saville, Gordon (General). "Interview with Lt. General Gordon Saville." Interview by Thomas A. Sturm, United States Air Force Oral History Program, Washington, D.C., March 26–29, 1973.

Schwartz, Jules. "The Development of JOVIAL." *Association for Computing Machinery SIGPLAN Notices*, Vol. 13, No. 8, August 1978, pp. 208–213.

Seife, Charles. *Decoding the Universe.* New York: The Penguin Group, 2006.

Severs, Hugh C. (Major). "The Controversy behind the Air Corps Tactical School's Strategic Bombardment Theory: An Analysis of the Bombardment vs. Pursuit Data between 1930–1939." Research Report No. 97-0126. Air Command and Staff College, Maxwell Air Force Base, Alabama, GA, 1997.

Shaffel, Kenneth. *The Emerging Shield: The Air Force and the Evolution of Continental Air Defense, 1945–1960.* Washington, D.C.: Office of Air Force History, United States Air Force, 1991.

Shalett, Sidney. "First Atomic Bomb Dropped on Japan." *New York Times,* August 6, 1945, p. 1.

Smullin, Louis. "Interview with Louis Smullin." Interview by Stuart Leslie, Massachusetts Institute of Technology, Cambridge, MA, January 20, 1988.

Snyder, Samuel S. "Influence of U.S. Cryptologic Organizations on the Digital Computer Industry." *Journal of Systems and Software,* Vol. 1, 1979, pp. 87–102.

Sprague, Richard E. "A Western View of Computer History." *Communications of the ACM,* Vol. 15, No. 7, 1972, pp. 660–689.

Stern, Nancy. *From ENIAC to UNIVAC: Appraisal of the Eckert-Mauchly Computers.* Boston: Digital Press, 1981.

Stern, Nancy. "The Eckert-Mauchly Computers." *Technology and Culture,* Vol. 23, No. 4, October 1982, pp. 105–111.

Stever, Guy. *In War and Peace: My Life in Science and Technology.* Washington, D.C.: Joseph Henry Press, 2002.

Stratton, Julius. "Memorandum to Members of the Research Laboratory of Electronics." MIT, Cambridge, MA, May 1, 1949.

Sturm, Thomas A. *The USAF Scientific Advisory Board: Its First Twenty Years, 1944–1964.* Washington, D.C.: USAF Historical Division Liaison Office, 1967.

Sugita, Yoneyuki. *Pitfall or Panacea: The Irony of U.S. Power in Occupied Japan 1945–1952.* London: Routledge, 2003.

Symington, Stuart. Testimony before the House Committee on Armed Services. In *The National Defense Program—Unification and Strategy,* United States Congress House Committee on Armed Servies, pp. 99–103. Washington, D.C.: U.S. Government Printing Office, 1949.

"U.S. Economy 1952." *Time Magazine.* November 17, 1952, p. 8.

"Outstanding Scientists." *Time Magazine.* June 7, 1954, p. 60.

Tropp, Henry et al. "A Perspective on SAGE: Discussion." *Annals of the History of Computing*, Vol. 5, No. 4, October 1983, pp. 392–401.

Truman, Harry S. Memo to James Forrestal, June 3, 1948.

U.S. Army Materiel Command. "Historical Focus: ENIAC—An Army Sponsored Revolution." Technical Report, U.S. Army Historical Office, Washington, D.C., July 28, 2000.

Underwood Computer Co. *Elecom 120* product brochure. Garden City, NY: Underwood Computer Co., 1953.

Valley, George E. "How the SAGE Development Began." *Annals of the History of Computing*, Vol. 7, No. 3, July 1985, pp. 196–226.

Vandenberg, Hoyt (General). Letter to Dr. James R. Killian. December 15, 1950.

Vandenberg, Hoyt, (General). Personal journal. August 24, 1948.

Vazquez, Donald 'Bud' (Lt. Col. USAF). "Building-to-Shelve Prototyping: Undercutting Doctrinal Development" Masters thesis, School of Advanced Air Power Studies, Air University, Maxwell AFB, Alabama, GA, August 1995.

Wallace, Henry A. "Report of the Secretary of Commerce." United States Department of Commerce, Washington, D.C., April 8, 1946, pp. 33-45.

Walterscheid, Edward C. "The Need for a Uniform Government Patent Policy: The D.O.E. Example." *Harvard Journal of Law & Technology, Vol.* 3, Spring 1990, pp. 109–113.

Ward, Jack. "The Transistor's Early History from 1947 to the 1960s." *Antique Radio Classified*, Vol. 15, No. 12, 1998.

Watson, Thomas J. (Jr.). *Father, Son, & Co.: My Life at IBM and Beyond*. New York: Bantam Books, 1990.

Weaver, Warren. "Science and Complexity." *American Scientist*, Vol. 36, 1948, pp. 537–539.

Weiss, Eric. "An Wang, 1920–1990." *Annals of the History of Computing, Vol.* 15, No. 1, 1993, pp. 60–69.

Wieser, C. Robert "From World War II Radar Systems to SAGE" (Speech at the Computer Museum. Boston, MA). *Computer Museum Report*, Vol. 22, Spring 1988, pp. 3–7.

Wieser, C. Robert. "Cape Cod System." *Annals of the History of Computing*, Vol. 7, No. 3, July 1985, pp. 363–371.

Wildes, Karl L. *Electrical Engineering at the Massachusetts Institute of Technology.* Unpublished manuscript, 1974. (Available at MIT Archives, Cambridge, MA.)

Wildes, Karl L. and Lindgren, Nilo A. *A Century of Electrical Engineering and Computer Science at MIT, 1882–1982.* Cambridge, MA: MIT Press, 1985.

Wilkes, Maurice V. "A Tribute to Presper Eckert." *Communications of the ACM*, Vol. 38, No. 9, September 1995, pp. 20–22.

Wilkes, Maurice V. "The Origins and Growth of Electronic Engineering—A Personal View." Speech at the Computer Laboratory, University of Cambridge, London, June 15, 2006.

Winkler, David F. *Searching the Skies: The Legacy of the United States Cold War Defense Radar Program.* Langley, VA: USAF Headquarters Air Combat Command, 1997.

Winstock, Jack and Gilbert, Willie. "Tom Corbett: Space Cadet." Script of TV series first broadcast 1950–1955. Produced by Rockhill Productions. (Available at http://www.solarguard.com/tcintro.htm.)

Wohlstetter, Albert and Rowen, Henry. "Objectives of the United States Military Posture." Technical report RM-2373. RAND Corporation, Santa Monica, CA, May 1, 1959.

Wolf, Marvin. "From Rocketeers to Solar Sailors." *Los Angeles Times Magazine.* November 14, 1999, pp. 33–36.

Wolk, Herman S. "The Founding of the Air Force." *Air Force: Journal of the Air Force Association, Vol.* 79, No. 9, September 1996, pp. 3–7.

Wong, James. "Vigilance and Vacuum Tubes: The SAGE System 1956-1963." Speech at the Computer Museum History Center, Mountain View, CA, May 19, 1998.

Zahl, Harold A. *Electrons Away or Tales of a Government Scientist.* New York: Vantage Press, 1968.

Zraket, Charles and Rose, Stanley. "The Impact of Command, Control, and Communications Technology on Air Warfare." *Air University Review*, Vol. XXIX, No. 1, November–December 1977, pp. 82–96.

Index